Research Report no. 89

Jerome O. Gefu

Pastoralist Perspectives in Nigeria

The Fulbe of Udubo Grazing Reserve

Nordiska Afrikainstitutet
(The Scandinavian Institute of African Studies)
Uppsala 1992

Indexing terms
Pastoralism
State
Nigeria
Fulbe
Nature reserves
Grazing

ISSN 0080-6714
ISBN 91-7106-324-2

Editing: Madi Gray

Printed in Sweden by Motala Grafiska, Motala 1992

Contents

Tables

Maps

Acknowledgements

This report is part of a larger socioeconomic survey of pastoral development alternatives in selected grazing reserves in Nigeria. The larger study was commissioned under the Second Livestock Development Project (SLDP), a World Bank livestock development assisted project in Nigeria. Major funding for field work came from the National Livestock Projects Division (NLPD) of the Federal Ministry of Agriculture and Natural Resources.

I wish to acknowledge the numerous individuals and institutions that, in one form or another, contributed to the production of this volume. The help and cooperation received from the district head and people of Gamawa and especially the numerous residents of Udubo is greatly appreciated and thankfully acknowledged. Field work was especially facilitated by the friendly and cordial atmosphere provided by Sarkin Udubo and his staff, who made the task of "rapport building" pleasurable.

Grateful acknowledgement is made for the logistic support received from Professor E.O. Oyedipe, Director of the National Animal Production Research Institute (NAPRI) at the Ahmadu Bello University in Zaria; Professor Saka Nuru (former Director of NAPRI); and Dr E.O. Otchere, leader of the Livestock Systems Research Programme at NAPRI. I also wish to thank our enumerators, the late Edwin Ogbole, David Samaila, Nuhu Sani and Bala Daniel for their dedication. Thanks go to research associates Hassan Ahmed and Hassan Umar for their assistance and long treks in the grazing reserve in search of our respondents. Similarly, my appreciation goes to the Ahmadu Bello University, Zaria for the continued support received while I was in Sweden.

Sincere appreciation goes to the Nordic governments and peoples for hosting me at the Nordiska Afrikainstitutet (NAI) in Uppsala as a Guest Researcher between December 1989 and April 1990. The months I spent at NAI afforded me the opportunity of writing the major part of this report. The resources—both human and material—made available to me at NAI, the University of Aarhus, the Centre for Development Studies in Copenhagen and the Chr. Michelsen Institute in Bergen contributed in no small measure to the completion of the report.

Colleagues in Uppsala offered critical and helpful comments on earlier drafts of the report. Among these are Anders Hjort af Ornäs, Director of NAI; M.A. Mohamed Salih, Jonathan Baker, Inge Tvedten, Kjell Havnevik, Margareta von Troil and Mai Palmberg. Secretariat and logistic assistance was received from Louise Simann, Ann-Marie Vintersved, Karl Eric Ericsson, Thomas Ridaeus, Birgitta Fahlander, Håkan Gidlöf, Ingrid Andersson, Susanne Linderos, Birgit Hegart, Annica and Niek van Gylswyk, Kent Eriksson and other support staff. The warmth, cooperation and kindness I received from these persons greatly facilitated the completion of this report. Madi Gray's painstaking editorial contributions are gratefully acknowledged.

Finally, the moral support received from my wife Tai and lovely daughters, 'Kemi and 'Femi, sustained me immensely.

All views expressed herein are those of the author and do not necessarily reflect the views of any of the aforementioned institutions and persons.

March 1992

Jerome O. Gefu

National Animal Production Research Institute
Ahmadu Bello University
Zaria, Nigeria

Acronyms and Expressions

ADP	Agricultural Development Project
Agoi	term for Garkayi
ara	leasehold
Ardo	head of a group of Fulbe families
aro	rent
Bororo	see M'bororo
boru	cattle disease, swelling mouths, abnormal salivation, cracked hooves
bugau	disease prevalent among sheep causing confusion and rapid death
daji	fallow land
dangere	*zornia glochidiata* , a legume causing bloating of cattle if grazed sole
dilali	middle-man, sells cattle and small animals
dusa	husks of grain
fadama	riverine, wet lowlands
gamba	a valued grass fodder
Garkayi	a nomadic clan of pastoralists
gishu	goat disease, prevalent among cross-breeds
GNP	gross national product
Hakimi	district head
harawa	crop residue
harbou	a shock to grazing cattle, causes collapse and possibly death
IRDP	Integrated Rural Development Project
jangali	cattle tax paid by Fulbe pastoralists
jihad	holy war (islamic)
jingina	mortgage
kindirimo	yoghurt
LDCs	less developed countries
LGA	local government authority
LIBC	Livestock Improvement and Breeding Centre
MACBAN	Miyetti Allah Cattle Breeders Association of Nigeria
madara	fresh whole milk

mai	butter
Mai	Anguwa ward head
mallam	a scholar of Islam, a scribe
M'bororo	derogatory term for Garkayi
naira,	kobo Nigerian money, abbreviated No.
NAPRI	National Animal Production Research Institute, Ahmadu Bello University in Zaria
NLPD	National Livestock Projects Division
nono	skim milk
RBDA	River Basin Development Authorities
ruga	homestead, compound
Sarkin	head, leader
Sarkin Fada	chief of security
shimilo	cattle disease, affects skin, eyes and acute diarrhoea, can be fatal in two days
SLDP	Second Livestock Development Project
TLU	Tropical Livestock Unit = 250 kg = 1 head of cattle, 5 sheep or 5 goats
Yabaji	term for Garkayi
zakkat	a tithe of one bull in every 30 bulls or one cow in every 40 cows owned

Grass and legume fodders

aldudure, bulude, chiasuwa, daramua, garlaba, jarugai, jemerai, jile, kajale, lablab, paguri, shinle, sohbe, stylo

Foreword

Nigeria has over seventy per cent of its people engaged in crop and/or animal agriculture, deriving their livelihood from agricultural production either directly or indirectly. Crops and livestock production are, therefore, a strategic socioeconomic activity in Nigeria, especially in the rural areas. Abundant natural resources exist to support a sustainable agricultural sector. Yet Nigeria continues to experience acute food shortages, especially animal source protein both in qualitative and quantitative terms. The situation has been further exacerbated by the unfavourable terms of trade and a drastic devaluation of Nigerian currency.

The worsening food supply—especially protein—has prompted a variety of programmes mounted by the different Nigerian governments to correct the situation through research and development.

For many years a number of agricultural policies and programmes (livestock inclusive) have been embarked upon by various administrations. Such Programmes as settlement of nomadic pastoralists and the establishment of grazing reserves as attempts to enhance livestock production have achieved little, either because they were based on concepts that are alien to the producers and the production system or because of unavailability of other support inputs such as water, animal health services, etc. Many livestock and pastoral programmes run into logistic problems and so the level of livestock productivity continues to be low even in the face of the high annual population growth rate (estimated at 3.5 per cent). This is an obvious indication that past agricultural thrust had not produced the desired results to meet the food requirements of an ever growing population.

It is in the face of these and other difficulties that it has become necessary to take a hard look at livestock development policies in an attempt to map out alternative and viable strategies for enhancing production especially at the producer's level. This is in the belief that Nigeria's livestock requirements can best be met by enhancing the productivity of millions of pastoralists who contribute about 80 per cent of the country's livestock needs. It is to this group, therefore, that aggressive and favourable livestock policies should be directed in order to boost local production of livestock.

In this book the author attempts to unveil the fundamentals of pastoral production so as to pave the way for meaningful state intervention in the present production systems. A meticulous and synthesized account of important aspects of pastoral producers, including aspects of their production organization is given against a backdrop of externally induced change among pastoralists. A vivid account of the household economy is undertaken to bring to light the complementarity and interdependence of crop and amimal agriculture. The book draws from field research among nomadic and agro-pastoral producers in Bauchi state, Nigeria, as well as on the wealth of experience of the author. Gaps and lapses in Nigeria's pastoral development efforts are highlighted. Options for effective and sustainable pastoral development agenda are proffered.

This study is most timely, especially at a period when animal agriculture is being re-awakened. It is sincerely hoped that this study will prove to be most valuable to students of agriculture, rural development, government and non-governmental organizations concerned with livestock and pastoral development, agricultural policy makers in governments, manufacturers and others with a serious interest in effecting positive and sustained development in Nigeria's livestock industry.

March 1992

E.O. Oyedipe

Professor and Director
National Animal Production Research Institute
Ahmadu Bello University, Zaria, Nigeria

1. Pastoralism in Transition

The importance and role of livestock in the Nigerian economy can be demonstrated by the large numbers and diverse species found. They are well adapted to the ecological conditions prevailing in different parts of the country. Livestock production forms the basis of the socio-cultural, economic and socio-political organization of over 9 million pastoralists. This group of livestock producers control the bulk of the nation's livestock population. Notable among them are the pastoral Fulbe who maintain over 85 per cent of Nigeria's livestock population. For these people, livestock breeding is the core of their socio-economic, cultural and political organization. Other non-pastoralists who raise small ruminants (grazing animals) and fatten limited numbers of cattle also derive a substantial portion of their income from keeping livestock.

Livestock production contributes significantly to Nigeria's national income. The livestock sub-sector provides employment opportunities for several million people in rural and urban areas either by their direct involvement in animal and/or crop production or through their employment in the various agro-allied industries. Livestock production is the source of about 40 per cent of that part of the national income which is derived from agricultural production. It provides about 58.5 per cent of the nation's meat consumption and contributed 7.4 million USD to the country's gross national product (GNP) in 1983 alone.

Despite the size of the Nigerian livestock population, serious deficiencies in local supplies of meat and meat products have been recognized (Federal Government of Nigeria, 1981; Olayide, 1976). A wide gap exists between the level of local production and national needs and demand. For instance, the total demand for meat in 1980 was estimated to be 388,990 tonnes, whereas the supply was 275,340 tonnes for the same period, a deficit of 113,650 tonnes. Similarly in 1981, there was an estimated deficit of 110,600 tonnes (Federal Government of Nigeria, 1981:104). It should, however, be noted that these figures only reflect the estimated demand for meat in comparison to local supplies. If the nutritional need—as against effective demand—is weighed against the supply, the deficit would be more alarming. The human/livestock ratio has steadily declined from 1:0.23 in 1960 to 1:0.17 in 1987 (Table 1). This deficit has continued to increase during the late 1980s and in the early 1990s.

The daily per capita protein intake of the average Nigerian falls "far short of officially estimated minimum requirements of 70 grams of total protein and 35 grams of animal protein per person per day" (Olayide and Olayemi, 1975:252). Only 8.4 grams of the total 53.8 grams of protein consumed by Nigerians is derived from animal sources. This suggests that the contribution of animal products to protein consumption is less than 16 per cent. The country is, therefore, not only a net importer of livestock and livestock products but also suffers from a daily per capita protein intake deficit. Similarly, the daily calorie supply per capita, estimated at 2146, (World Bank, 1989:218) is well below world average figures.

Table 1. *Human/livestock ratio in Nigeria (1960-1987)*

Year	Human population[a] (million)	Cattle[b]	Sheep[b]	Goats[b]	TLU[c]	Human/livestock ratio
1960	42.4	6.5	2.7	13.3	9.7	1:0.23
1970	54.3	8.0	4.7	16.3	12.2	1:0.22
1972	59.8	8.2	5.4	17.1	12.7	1:0.21
1976	67.8	9.0	6.9	18.6	14.1	1:0.21
1980	77.1	10.0	8.9	20.3	15.8	1:0.20
1985	91.2	11.1	12.2	22.7	18.1	1:0.20
1987	116.2	12.0	13.8	23.7	19.5	1:0.17

[a] Based on figures from U.S. Department of Commerce, 1983, and estimates from Economic Intelligence Unit (EIU), 1990. Nigeria: Country Profile (1990-91) Annual Survey of Political and Economic Background, London.

[b] Estimated livestock population, FLD, 1987.

[c] TLU = Tropical Livestock Unit = 250 kg = 1 head of cattle, 5 sheep or 5 goats.

The deficit in the local supply of meat has forced the Nigerian government to take a number of steps to alleviate shortages. One such measure has been the importation of livestock, meat and meat products as a short-term measure to combat shortages. These huge imports constitute an attempt to supplement local sources of protein.

Local deficits in protein intake, coupled with huge import bills and the corresponding drain on foreign reserves calls for a critical appraisal of the policies involved in the development of the livestock sub-sector.

The number and variety of livestock in Nigeria as well as the abundance of natural pasture offer unique opportunities for dealing with local shortages. Any effort to enhance local production would contribute toward making Nigeria self-reliant and self-sufficient in meat and livestock production.

There is a dearth of information about the country's livestock population and herd structure. Nigeria's livestock population has, however, been estimated to be between 12-16 million cattle, some 13.5 million sheep, about 26 million goats, approximately 2.2 million pigs and 150 million poultry (local and exotic). Regretably, despite the size and diversity of its livestock resources Nigeria has not been able to provide the average citizen with the minimum requirements of animal-source protein. Many reasons have been advanced for the poor capacity of locally available livestock to meet domestic needs. These range from claims associated with the producer's "social-psychological enclave" to institutional and policy factors.

The problem of producing sufficient livestock and animal products locally grew at the attainment of political independence and worsened as the years went by. Importation has been used to compensate for the short supply of meat and animal products. By 1976 the volume of food and live animals imported into the country was larger than exports of the same commodities. As local supplies dwindled, increasing foreign exchange allocations continued to be made for the procurement of meat and livestock products.

Colossal import bills were incurred by both federal and state governments by early 1980s. By 1983, for example, Nigeria had a record import bill of about 1,121 million naira (1 naira = 1.30 USD at the then exchange rate) for food and live animals alone. In 1984, a total of 549,768 head of cattle were imported (Federal Livestock Department, 1984). The figure represents about 43 per cent of the total supply in the country and at that time was valued at 429.40 million naira, at an annual average price of 781 naira per head.

Faced with the necessity of meeting the protein requirements of its growing population, Nigeria introduced a number of policies and programmes to improve and boost local production. Government intervention in the indigenous livestock sector was based on certain assumptions which date back to the colonial era and probably beyond.

The Nigerian livestock sub-sector is often said to be a sector of unfulfilled potential because its potential and actual contribution to the GNP as well as to nutrition could be greatly enhanced if the vast number of "traditional" producers could be persuaded to practice "modern" methods of livestock production. The thinking is that a higher turnover and improved meat quality would be attained if and when the sub-sector is modernized and improved management practices are adopted by Nigeria's pastoral producers.

Pastoral producers have been blamed for low productivity of local livestock as a result of resistance to change and the reluctance of pastoralists to adopt innovations. This thinking may be grossly misleading when any attempt is made to understand the fundamentals of pastoral production in Nigeria and indeed, those of many other pastoral systems in the developing world. Critics of the traditional production systems have not addressed the main source of the complex of pastoral problems, rather they have found easy targets to blame for the low productive capacity of Nigeria's livestock.

What needs be seriously examined by critics of pastoralism seems to be the quality and number of existing innovations that pastoralists can readily use. Alternatively, how appropriate are the research findings on improved livestock production to the varying conditions of pastoralist production? The question should be posed about whether improved livestock production packages are available in the right form and at the appropriate time to the pastoralists and yet are rejected without good reason.

There is no doubt that government attempts to bring about change in pastoral production systems. Some of the major problems of the livestock sub-sector, however, derive from these very attempts to improve the systems through the imposition of inappropriate practices (Gefu, 1990).

Intervention rarely takes into account the existing systems into which change is being introduced. In most cases, the changes introduced are economically unrealistic even for those with political and economic power, such as modern ranchers and feedlot operators. Traditional pastoralists are in a still more disadvantageous socio-economic and political position when it comes to utilizing advanced technology for animal production. Where innovations or the changes introduced are feasible and affordable, pastoralists are known to accept them readily.

Pastoralists and Pastoralism

Among the major factors responsible for the declining performance of the livestock sub-sector are socio-economic and sociocultural variables, inconsistent and unsustained government policies, credit, land tenure and land rights and institutions, marketing and other infrastructural constraints (Federal Government of Nigeria, 1981; Gefu, 1986; Winrock International, 1978; Simpson and Evangelou, 1984). The phenomenal crisis in which the Nigerian livestock industry is engulfed is part of the larger food and hunger crisis that has ravaged so many developing countries in the past two decades.

Attempts made in the past to come to grips with critical food shortages, especially of animal-source protein, have led Nigeria to set up several special study groups to provide an answer to the problems facing the livestock industry. Many of these efforts do not seem to have yielded appreciable results.

Part of the difficulty in grappling with the basic dilemma of livestock producers, particularly the numerous "traditional" producers, can be associated with the approach adopted in the past. The problems facing Nigeria's livestock industry have until recently been conceptualized and examined in a narrow disciplinary fashion. The need for an interdisciplinary approach to the study of livestock and pastoral production has, however, been underscored in recent times by institutions and organizations concerned with the problem.

Recognising the potential merits of an interdisciplinary approach to the problems of livestock development in Nigera, the National Livestock Projects Department, in collaboration with the World Bank, convened a meeting of specialists on important aspects of livestock production. The objective was to bring different dimensions of disciplinary emphasis to bear on the diverse problems of pastoral production in order to better appreciate constraints and work out modalities for containing or ameliorating production bottlenecks. The starting point was an evaluation of existing grazing resources and the development of alternatives for improved range utilization of the existing grazing reserves.

The idea of establishing grazing reserves in Nigeria was first proposed by the World Bank study of 1949-1954 and was officially adopted by the Federal Government during the 1970-74 plan period. The underlying motive behind grazing reserves is to provide pasture, water and animal health facilities for the use of pastoral producers and consequently to induce pastoralists, especially the nomadic group, to settle.

An offshoot of the grazing reserve concept, the agro-pastoralist model, was recently adopted by the Nigerian government as one of its strategies for pastoral and livestock development. It seeks to bring pastoralists into the core of livestock development by turning producers into modern entrepreneurs. If producers take full responsibility for managing their own resources—especially pasture and water—it is anticipated that some of the pitfalls of the grazing reserve strategy will be averted.

A Multidisciplinary Study

The objective of the present study is twofold. First, it aims to synthesize and critically evaluate pastoral development programmes in Nigeria from a historical perspective, drawing parallels with other African and Asian pastoral societies. Secondly, it aims to offer a descriptive analysis of a concrete pastoral development programme, focusing on the important socio-economic impact and the need for development intervention in pastoral communities.

To this end specific variables were investigated. These include the types of livestock, herd size and ownership structure; stock management systems and use of crop residues; present land tenure and agricultural practices; the nature of future interaction between project beneficiaries and non-beneficiaries. Also examined were the perceived needs and wants of livestock owners in terms of their wish to settle: if, why, how and where; pastoralists' requirements in terms of grazing, water, veterinary care and marketing facilities; their willingness to pay for infrastructural developments and needs or desire for credit; their preparedness to accept control of stock numbers when necessary; their acceptance of development proposals in terms of the agro-pastoralist model as well as their requirements for social services such as health and education. Thus the study examines the perception of pastoral producers who are potential beneficiaries of the agro-pastoralist pro ject.

This report is part of a larger multidisciplinary study of pastoralists in Northern Nigeria, carried out between November 1986 and August 1988 in the Gamawa Local Government Area of Bauchi State, Nigeria. The author served as one of the consultants assembled to examine specific aspects of grazing reserve development in Nigeria.

A grazing reserve established at Udubo serves different pastoral groups in the area. The observations in this report are based on data generated by the author through interviewing 201 pastoral and non-pastoral households that have been utilizing the grazing reserve for several years.

A descriptive analysis such as the one attempted here is important to gain a fuller understanding of the various interacting socioeconomic, cultural and political factors that affect pastoral production. Such information is required because of the numerous state-led interventions that have been embarked upon since colonial times.

The Survey

The survey utilized formal and informal interview schedules along with participant observation. Questionnaires were administered by trained and experienced enumerators. Sampling was partly stratified by major occupational activity. One group was primarily engaged in pastoral production—the pastoral Fulbe—while the other was primarily occupied by raising crops which they supplemented with stock raising—the Hausa cultivators.

The limited time available for the study allowed for 201 households from both groups to be selected and interviewed. The sampled population included 171 Fulbe and 30 Hausa households. The unit of analysis was the household, hence only heads of households were interviewed.

In a survey of this nature, it is often difficult to gain access to respondents to obtain valid information concerning their production activities. Getting a response may be as troublesome as locating and keeping appointments with the respondents. The services of the local authority (district head) were solicited for two main reasons: to authenticate our presence and to locate respondents with relative ease. The pastoral Fulbe respondents were spread out in numerous compounds or *ruga* and were often gone to market or in their fields harvesting crops during a major period of the study. After consulting with the *Hakimi* (district head) and the *Sarkin Fada* (chief of security), the heads of groups of Fulbe families, *Ardos,* were employed to lead our party to all Fulbe locations and several other points on the grazing reserve. Meetings were arranged and held first with all the *Ardos* and later with individual *Ardos* in the company of Fulbe household heads.

We spent the first few days on building *rapport* and becoming familiar with the grazing reserve. No interview schedule was administered until our intentions were understood reasonably well. The total number of Fulbe interviewed represents approximately 42 per cent of the estimated Fulbe population in and around the reserve.

A sample of 30 Hausa respondents was selected on the basis of their livestock raising activities and the convenience of reaching them.

The survey was conducted during two periods, the wet and dry seasons. The dry season survey targeted sedentary pastoralists and agriculturalists, while the wet season survey concentrated on the nomadic pastoralists who are usually not found on the grazing reserve in the dry season. Conditions prevailing during these two periods are described in Chapters Three and Four, where areas of convergence and divergence relating to pastoral activities are highlighted.

2. Perspectives on Pastoralism in Nigeria

There is no general consensus as to the origin of the pastoral Fulbe who live in Nigeria today. Accounts of the exact period of their arrival and settlement in various locations in Hausaland are not consistent. An aspect common to accounts of their origin, however, traces the pastoral Fulbe to the Senegambia region from where they spread eastwards. They have generally been associated with the Western Sudanese socio-political and cultural milieu. Some Fulbe legends trace their descent to Ukbatu, an Arab who married an African called Bajjomanga hundreds of years ago.

Puzzling as the origins of the Fulbe people may be, their language (Fulfulde) is similar to Wolof, which indicates that the Fulbe belong to a linguistic group which closely resembles the Niger/Congo category. Because of this linguistic resemblance, it has frequently been suggested that Fulbe communities found in different parts of sub-Saharan Africa derive from the Senegal river basin (Stenning, 1959).

The pastoral Fulbe, as an ethnic group, first appeared in Hausaland in the reign of Sarkin Kano Yakubu between 1452 and 1463 AD and soon became a significant minority in the population of the Hausa kingdoms. Many became prominent Islamic scholars and judges and contributed significantly to the spread of Islam to communities in West Africa. While some gave up animal husbandry as a primary occupation, most continued to herd animals full time.

Soon, their herds of cattle became a source of revenue which the Hausa rulers eyed covetously and began to tax greedily (Wall, 1988). The imposition of cattle tax on the Fulbe saw the advent of the incorporation of pastoralists into the larger socio-political and economic arena of the state, which soon became a source of friction between the pastoralists and their Hausa overlords.

Apart from their formal relationship with the local authorities, pastoralists have long informal relationships with other communities which largely constitute crop producing groups. The link to these agricultural communities is symbiotic in nature. Even though the Fulbe diet consists largely of milk and milk products they rely on non-pastoral farmers and traders for a variety of items of food and other consumables.

Contrary to some opinions, the Fulbe do not rely exclusively on the milk and milk products derived from their herds. Complete dependence on the consumption of milk would not be possible due to the seasonal production of milk in the savannah where the pastoralists are predominantly found (Waters-Bayer, 1988).

Furthermore, the cow/human ratio must be high if enough milk is to be produced to meet nutritional requirements all year round. Dahl and Hjort (1976) calculated a cow/human ratio of over 100:1 to make it possible to be completely dependent on milk as a food. The Fulbe, therefore, exchange portions of their unconsumed milk products for cereals which provide higher calorific values. The pastoral Fulbe of old are known to have institutionalized food-exchange relations with sedentary populations whereever contact was made (Burnham, 1980).

As the empires of Hausaland grew, so did the expropriation of the surpluses produced by the citizens. Hausa cattle owners were made to pay a tithe or *zakkat* of one bull in every thirty, or one cow in every 40, while the Fulbe paid tax, *jangali*, on their cattle herds. It is not known how much was levied, but the tax varied with the size of the herd owned and managed.

The imposition and collection of *jangali* continued up to the early part of the 19th century when Usman dan Fodiyo—a Fulbe islamic scholar—led a holy war, *jihad*, against the Hausa overlords. Pastoral communities all over Hausaland rallied, providing a large following to support their kinsman and, more importantly, in anticipation of socio-economic and political leverage and protection. The pastoralists hoped that under the rule of a person like Usman dan Fodiyo, their tax burden would be lightened and perhaps abolished. Furthermore, they expected their productive base—pasturage and water—would be guaranteed and possibly enhanced.

During the *jihad* period (1804-1830) some wealthier pastoralists with an inclination for urban life and more attached to the administration of the day, opted for a sedentary life, using prisoners of war as slave labour for duties like animal herding. Furthermore, the frequent raids carried out on the herds of the more mobile pastoralists compelled some to live within relatively safer walled settlements. The majority of pastoralists, however, continued to move from one place to another with their animals.

The outbreak of rinderpest that occurred between 1889-1893 adversely affected pastoralists' herds. The drastic reduction of the herd population made it unnecessary for many pastoralists to continue using a migratory pattern of animal husbandry. Furthermore, uncertainty of how the contagious virus disease was spread and the severity of the outbreak elsewhere compelled many pastoralists to consider settling and supplementing their pastoral activities by crop farming.

The pacification of Northern Nigeria by the British Protectorate at the turn of the century led to the reduction of inter-ethnic conflict and insecurity about lives and property. The relatively peaceful and tension-free atmosphere, coupled with the prohibition of slavery, resulted in the resumption of migratory pastoralism by some pastoralists. Many Fulbe who had relied on slaves were deprived of this source of labour.

Moreover, the relative success of campaigns in controlling rinderpest, helped to rebuild pastoral stock. As herds grew, so did the farming activities of agricultural communities. More land was cultivated, leaving less pasture land for grazing animals. Population growth put increasing pressure on the available land as former pastures were farmed. Riverine areas previously used for grazing in the dry season were increasingly devoted to cash crop cultivation to meet British and Nigerian urban demands.

The cattle tax, *jangali*, was raised, in some places by almost 100 per cent. It was reported that the Emir of Kano increased *jangali* to 5,000 cowries from 2,500 per head after the great rinderpest epidemic (Hill, 1977). All these factors compelled pastoralists to resume the migratory pattern of animal herding. Under the conditions prevailing, the pastoralist can be seen as reacting in accordance with the dictates of the socio-political and economic environment.

Over time, the relationship of pastoralists to the state continues to be manipulated to serve the purposes of the operators of the state apparatus. The incorporation of pastoral producers into the economic realm of the state is, therefore, not a new phenomenon. Successive administrations of pastoral societies the world over continue to play a dominant and active role in how primary producers organize production.

It is against this general background of Nigeria's pastoral socio-economic and political history that we shall examine how some of the activities of the state influence pastoral producers in the country.

Pastoralists and the State

In most developing countries intervening agencies held that agricultural development was an area that required transformation from simple subsistence production to a market-orientation. Our examination of the different reasons for state intervention in the activities of primary producers gives credence to this notion. The drive for the development of commodity relations in primary production dates back to the colonial period with the incorporation of pastoral producers into the market economy.

A review of state intervention in pastoral production shows three main categories of interventions that correspond approximately to three different historical periods.

The first period, the era of colonialism, saw externally induced change in the production activities of pastoralists, primarily in the provision of infrastructure and supplies. The second period, from the time of political independence up to the oil boom, is marked by interventions intended to commercialise livestock production through attempts to replicate western systems of livestock production. The third period commenced with the worsening debt burden suffered by most primary producing countries and was characterised by the reorganization of pastoral production systems.

This broad classification is not, however, mutually exclusive as frequently aspects of one period are found in other periods. The classification scheme offered here is only meant to assist us in analysing the interventionist role of the state from a historical viewpoint.

Prior to 1900 the socio-economic and political life of pastoralists did not undergo the kind of changes experienced under colonial domination. Pastoralists were in conflict with communities far removed from their immediate environment. Raids on their herds were constantly launched as inter-ethnic conflicts deepened (Hopen, 1958), which adversely affected pastoral activities. The situation was worsened by the widespread rinderpest epidemic of 1887-1893 and aggravated by the famine of 1913-1914.

Some of the more important changes that occurred between the colonial era and the present are highlighted below.

Pastoralists and the Colonial State

The colonial era in Nigeria dates from the establishment of the Protectorate in 1900 to the attainment of political independence in 1960. The British colonial administration had taken at least indirect control of the area inhabited by the pastoral Fulbe by 1903. At this time the ravages of the rinderpest epidemic were still evident, so one of the first concerns of the colonial administration was to bring bovine diseases under effective control. The colonial government initiated disease control programmes to raise livestock production and ultimately generate a higher income for itself through the imposition and collection of a cattle tax.

The pastoral Fulbe were consequently drawn into the market economy as they sold their stock for "modern currency" to pay taxes and meet the demands of modern times. The introduction of cash seems to have forced the pastoralists to raise more stock than they needed, so they could pay their taxes. As higher cattle tax was levied, more animals had to be raised and sold to meet government demands.

The suspicion may not be unfounded that pastoralists were assessed at a relatively higher tax rate during the earlier periods of colonial domination. This is due to the ease with which a tax assessment can be done on the basis of the size of a herd. It was relatively more difficult to do a tax assessment of agricultural produce than on the number of animals owned. The imposition of the cattle tax must have made it necessary for pastoralists to expand their production, while the government derived most of the benefits.

The increased tax burden impossed by the colonial state on pastoral producers was coupled with conflation due to poor terms of trade, as the colonial administration absorbed the very profits it promoted (Watt, 1983). This further facilitated the appropriation of a large economic surplus from pastoralists, which was often deployed in bureaucratically directed economic activity in the name of promoting social and economic development (Saul, 1983).

In 1909, veterinary services were established within the newly constituted Agricultural Department. The veterinary services offered included both curative and preventive care, through campaigns against disease as well as through the control and treatment of outbreaks. Vaccination and dipping centres were established for the treatment and care of animals and for the prevention of serious bovine diseases. Initially, the services were provided at minimal or no cost to the pastoralist. In later years, small fees began to be charged for certain veterinary services.

Realizing the importance of the livestock sub-sector, the colonial administration embarked upon the establishment of Farm Centres. These centres provided feed supplements, mineral and veterinary supplies. They also offered advice on how to improve the production of local livestock. Farm Centres were established in various parts of Northern Nigeria between 1921 and 1934.

Stock breeding programmes were also started. Livestock improvement was sought through the selection of quality local breeds and by cross-breeding exotic animals with indigenous ones. Emphasis was put on cross-breeding as it was thought that the characteristics of the exotic breeds—which were considered more productive than the indigenous breeds—would be passed on to the cross-breeds. The issue of the adaptability of both the exotic and the cross-breeds to the environmental and climatic conditions did not seem to be of prime interest at the time the cross-breeding programmes were initiated.

Following the recognition given to the potential role of cross-breeding, the Shika Stock Farm was established in 1927 to provide improved breeds for the Farm Centres. The rationale for establishing the Farm Centres was an attempt to provide models to persuade pastoralists, especially nomads, to settle permanently.

Little success was, however, recorded by either the veterinary services or the Farm Centres, as the intended beneficiaries of these programmes refused to "take advantage" of the services (Awogbade, 1982).

In the early 1900s the authorities saw a need to sedentarize nomadic pastoralists. Dr. Growers, then the Resident of Sokoto Province in the Northwest, argued that the future of livestock in Nigeria hinged on the sedentarization of the pastoralists. It was not until 1942, however, that the idea of settling nomads was actually put to work in the Jos area.

Jos is a plateau with rich grassland and soil supplied with abundant all-year river and underground sources of water. The area was also tsetse free. Each (pastoral) producing unit or household was allocated four hectares of pasture land in the area in the hope that pastoralists would not only settle permanently but would also engage in mixed farming. At this time, a milk processing plant was operating in Vom (near Jos). The plant was built in anticipation of a regular supply of raw milk from the herds of the pastoralists. The anticipated supply of milk did not materialize. The allocated patches of land were soon taken over by mining companies that sprang up following the discovery of tin in the Jos area.

This was one of the earliest attempts made by the colonial administration to settle pastoralists. Attempts continued, but met with resistance from the pastoralists (Dunbar, 1970). The pastoralists seem to have anticipated that undesired changes in their economic and social organization would follow sedentarization.

One of the terms of reference of the expert study group of the International Bank for Reconstruction and Development (World Bank) between 1949 and 1954 was to explore the difficulties faced by the colonial administration in their attempts to settle pastoralists.

At the time of the World Bank study, conventional wisdom held that the free-range management system used by the pastoralist was counter-productive, wasteful (i.e. resource-depleting) and undesirable. People felt that all possible means should be employed to discontinue it. The World Bank study not only reiterated the assumptions about pastoralists made by the colonial administration, but also made concrete proposals for settling them.

The World Bank team recommended the establishment of such facilities as marketing channels, watering points, veterinary posts and, most significantly, the establishment of grazing reserves in major producing areas of Northern Nigeria. In recommending the establishment of grazing reserves, the Bank argued that reserves would make it easier to provide the pastoralists with a social infrastructure as well as to control and treat bovine diseases. In this way, the pastoralists' basic needs would be better served, it was contended.

It was felt that efforts to improve livestock would be enhanced (eg. stock upgrading, cross-breeding with exotic animals, restocking, etc.). The general idea behind the grazing reserve programme was that better management practices would be used under conditions favourable to livestock production. There would be feeds and fodders, a regular mineral and water supply and veterinary services would also be readily available to enhance animal health and productivity.

The recommendations of the World Bank team were accepted and began to be implemented shortly after independence with the establishment of the Ruma-Kukar-Jangarai grazing reserve. Since then, the concept of grazing reserves has been consistently applied to solve the problems of pastoral and livestock production in the country.

It is evident that the assumptions held by the colonial administration about pastoralism were passed on to the independent Nigerian administration. The new state shared the same notion that pastoralism is inefficient and needed to be radically altered. Consequently, the development programmes embarked upon by the post-colonial Nigerian administration reflected the already existing antagonistic attitude towards traditional pastoral producers.

Between the 1930s and the 1950s the colonial government embarked upon several stock-raising schemes in Northern Nigeria, but

> ... none was mounted on a sufficiently large scale to make any drastic or far-reaching changes in the cattle situation. Some small "multiplication centres" were started with the purpose of upgrading stock, but the scale of operations was so small that effects were not felt beyond a limited range (Dunbar, 1970:122).

These multiplication centres still exist to date under the name Livestock Improvement and Breeding Centres (LIBC). These LIBCs have a variety of breeds of sheep, goat and cattle which are bred for improved progeny. These improved breeds are intended for distribution among small stock holders, especially those that live around the LIBC.

One of the author's visits to a typical LIBC revealed that claims far exceed the breed improvement that is actually achieved. Not only is breeding and animal conditioning minimal, the distribution of "improved breeds" to local farmers is almost non-existent. Moreover, inspection of records and observation of the physical condition of the animals showed morbidity and mortality rates to be high. This is probably a major reason for the apparently limited success of the Livestock Improvement and Breeding Centre project.

We have thus far focused our discussion on the involvement of the public sector in pastoral production activities. We shall now briefly examine the part played by the private sector in the development of the livestock industry.

Private Sector Involvement

The first attempt to establish commercial ranches was made in 1912 when the traditional ruler of Bornu (in the northeastern part of Nigeria) transported some cattle and sheep to Lagos (in the south) by rail. He made a substantial profit of about 90 per cent (Dunbar, 1970).

Numerous requests were received by the then Governor of the Protectorates of Northern and Southern Nigeria, Frederick Lugard, for permission to establish large-scale commercial ranching schemes. These requests were generally for very large expanses of land and usually proposed to operate the ranches in the Australian or American style. For example, in 1913, a request was made for "not less than 2,400 square miles of grazing land... within say 150 miles of present or projected government railways" (Dunbar, 1970:104-5).

Approval for ranches on such huge land areas was not given by Lugard. His decision may have been influenced by the fear that the colonial administration would lose control of certain land areas if several requests were granted. Nor did Lugard want to defy native laws which prohibited the sale of large pieces of property to expatriates. The latter reason was particularly important because Lugard's experience in India and East Africa suggested that large-scale European enterprises, especially in East and South Africa, had resulted in widespread disapproval and unrest. The colonial government did not want to risk political upheaval in Nigeria by allowing large-scale commercial enterprises.

Commercial ranching in the real sense of the word began in 1914 with the establishment of African Ranches Limited. However, the company was forced out of business and subsequently sold its assets to the colonial government in 1923. The performance of this ranch was not superior to that of the pastoral Fulani. For example, in 1920, the colonial government visited the ranch and reported that although the "cattle were in very good condition, ... they did not differ materially from the native-owned herds... The ranch cattle were no more protected from rinderpest and other diseases than were native cattle" (Dunbar, 1970:117).

Many of the pastoral policies and programmes conceptualized by the colonial administration were inherited and closely followed by the independent administration. The commitment of the new government to the pastoral programmes started by the colonial administration is evident in the various National Development Plans (see, for example, Federal Ministry of Information, 1962; Federal Ministry of Economic Development, 1981).

Pastoral programmes and livestock projects can be said to have been conceived and developed outside their Nigerian context. Often support has come from external interest groups. The push for the "modernization" of pastoral production was often accompanied by externally funded programmes and projects which had implications that were either unanticipated or counter-productive.

Grazing reserve programmes were launched in order to offer incentives to nomadic pastoralists to settle permanently. Yet little or no permanent settlement seems to have taken place (See Horowitz, 1979). The often cited example of the advantages over pastoralism derivable from the ranch system is the ease of control of diseases and predators in ranch animals. An outbreak of an epidemic could, however, be more devastating to a large flock on a ranch than it would be to nomadic herds.

Herds of nomadic pastoralists could move more quickly in response to an epidemic, or their movements could be planned so as to avoid disease-prone areas. As Dunbar pointed out, the animals on the ranch tended to be no more protected from disease than those of pastoralists.

Attempts made by both public and private concerns to "modernize" livestock production have met with varying degrees of success and failure. Those strategies, policies and programmes, which the post-colonial government inherited from their predecessors, are the subject of the section that follows.

Pastoralists and the Nation State

Following the attainment of political independence in October 1960, the new administration expressed no immediate concern about pastoral production or nomadic pastoralism. The nation state was preoccupied with its commitment to rapid economic development in general terms. This was clearly reflected in the 1962-68 First National Development Plan, which had certain macro-economic goals as its major objectives:

> To surpass the present growth rate of the economy of 3.9 per cent per year to achieve a rate of 4.0 per cent per annum and if possible, to increase this rate;
>
> To achieve this aim by investing 15 per cent of the GDP and at the same time endeavouring to raise per capita consumption by about 1 per cent per annum;
>
> To achieve self-sustaining growth not later than the end of the third or fourth national plan;
>
> To develop as rapidly as possible opportunities in education, health and employment; and to improve access of all citizens to these opportunities;
>
> To achieve a modernized economy consistent with democratic, political

and social aspirations of the people in a manner condusive to a more equitable distribution of income both among people and among regions; and

To maintain a reasonable measure of stability through appropriate fiscal and monetary policies (Federal Ministry of Information, 1962:23).

The above statement of development objectives does not make any clear-cut reference to a specific agricultural programme nor is there any discussion of pastoralism.

At independence, however, the World Bank report was reviewed and adopted as a consolidated programme called the "Fulani Amenities Programme Proposals". This programme proposed to offer, among other things, range improvement, fodder conservation for dry season feeding, improved pasturage, water development and the provision of supplementary feeds. The cost of this proposal was put at 3.5 million pounds sterling and responsibility for its execution was divided.

Development corporations in Nigeria and their agencies were to look after marketing, purchases of local stocks and fattening; while the Ministry of Animal and Forest Resources was responsible for other matters related to cattle development as well as veterinary services (Awogbade, 1982). A bill that legalised this proposal was passed in 1965 and called the "Grazing Reserve Law". The intent of the law has been described by Awogbade (1982:13) as:

> ... to enable the Ministry of Animal and Forest resources as well as the native authorities to create grazing reserves, so that the grazing rights of cattle owner can be fully protected by law. Besides, it was intended to reduce the constant friction between cattle owners and farmers. Provisions were made for pasture development and permanent water supplies. Another incentive which the law provided was to encourage gradual sedentarization of nomadic Fulani... Economic benefits in terms of regular marketing of their surplus products were to be provided. And where feasible, they would be allowed to form cooperatives. It was in pursuance of the above, therefore, that two forest reserves in Ruma and Jangarai... were merged into one to form the nucleus of what is now Ruma-Kukar-Jangarai grazing reserve.

The 1960s saw a number of projects in Northern Nigeria developed in collaboration with foreign agencies, among which was the United States Agency for International Development (USAID).

> Several range management demonstration sites have been opened in former forest reserves in Katsina, Sokoto and Plateau provinces. These are attempts to settle the Fulani and to introduce them to advanced practices. They are veritable "community ranches" but there is probably not space enough in Northern Nigeria to accommodate all pastoral Fulbe in this way (Dunbar, 1970:122).

The limited success of the attempts of the Nigerian government to settle nomadic pastoralists is a clear indication of the seriousness always posed by the land tenure issue. The land holding and land utilization pattern two decades ago was such that grazing land was more readily available than at the present time. The pressure put on land by industrial, agricultural and urban growth is even greater today, due to the many competing uses to which land can be put.

Immediately following independence several ranches were started in collaboration with international agencies. Three of them—two funded by USAID—will be mentioned here. Dunbar (1970:122) has succinctly described these projects:

> The Bornu Ranch, one of the AID projects, is a breeding ranch twenty-two miles southeast of Maiduguri in the Gombole Forest Reserve. It was begun in 1963 and complete Nigerian control was planned at the termination of the original Six-Year Plan in 1968. By 1967, 20,000 acres had been fenced and 300 Wadara cows had been purchased. An eventual herd of 1,000 breeding cows is planned and this would mean a total herd of about 4,000, including bulls and young. The cattle would be supported not only by natural forage but also by improved pastures and cultivated fodder crops.
>
> The other AID-supported ranch is the Manchok "fattening ranch" of 5,820 acres sited in a previously unused area just under the western scarp face of the Jos Plateau. It was also started in 1963. Although sited in a tsetse area, the ranch can be cleared and kept free of tsetse flies. The grass resource has not been diminished by severe over-grazing, as has much of the Jos Plateau. The Plateau cannot only be counted on to supply cattle to the ranch for fattening, but it provides cottonseed for feed as well. In 1967 there were about 560 head of cattle on the ranch and an actual annual production figure of 5,000 is planned. Cattle are removed by road and rail to Kaduna...
>
> The German venture, the Mokwa Ranch, is not only a "fattening ranch" but a research station as well. This ranch was established in 1964 on lands which were involved in the ill-fated Niger Agricultural Project in the

> early 1950s... In 1967 there were 600 cattle on this ranch of approximately ten square miles and an eventual annual production figure of 5,000 is envisioned.

Foreign interests in the programmes of the Nigerian government continued to grow. These foreign interests were documented in government statements of policy or development plans. All projects embarked upon after World War II were state-supported, with foreign agencies contributing part of the capital.

In the years of the second, third and fourth National Development Plans—1970 through 1985—similar statements were made concerning pastoral and livestock production. The primary objective of government had always been to increase livestock production and thereby make more beef and dairy products available to the Nigerian population, particularly to urban consumers.

The concern of the government has typically been expressed in words such as "the eradication of tsetse fly, the control of trypanosomiasis and other diseases and the settlement of nomadic herdsmen" (Federal Government of Nigeria, 1970:121). More specifically, the government has identified a factor militating against increased and improved livestock production in the country:

> Among the most serious deterrents to the breeding and improvement of cattle is the presence of tsetse fly in the southern well-watered parts of the country. The majority of cattle are therefore concentrated in the seasonally dry and less humid belts across the extreme northern part of the country where conditions are too arid for tsetse fly. The herdsmen are, however, forced to migrate in search of water and better pastures for their animals during the dry season and this nomadic way of life is not conducive to increased productivity and general livestock improvement. (Federal Government of Nigeria, 1970:121).

The removal of this bottleneck seemed to have been the main preoccupation of policy makers and major areas of policy were developed to tackle the problem:

> establishment of large-scale feed depots and livestock multiplication farms for the production of parent stock;
>
> subsidization of livestock inputs such as feeds, breeding stock, vaccines, drugs, equipment, etc. to livestock products;

encouragement of private ranching for beef, dairy, sheep and goat through the provision of improved pastures and fodder facilities for grazing, improved breeding stock and settlement schemes for the nomadic herdsmen;

intensification of veterinary and livestock production extension services... (Federal Government of Nigeria, 1981:106).

Of all the proposed government activities in the livestock sub-sector, the establishment of grazing reserves seems to have been emphasized the most. For example, the Third National Development Plan of 1975-80 proposed allocating a total of 22 million hectares to grazing reserves.

By the end of 1977, only 2 million hectares had been acquired by the State and Federal governments together. A 1978 report on pastoral nomads in Kwara State, submitted to the Federal Livestock Department, suggested that the grazing reserves planned by the state be established without further delay and that nomadic pastoralists be settled around villages (Federal Livestock Department, 1978).

Intervention aimed at reorganizing the production systems of pastoralists as well as infusing them with a sense of commercialism continued into the late 1970s and the 1980s. The First and Second Livestock Development Projects, jointly financed by the Nigerian government and the World Bank, were major landmarks in the country's livestock sub-sector.

The Philosphical and Political Basis for Intervention

The instruments which the state utilizes to impact on pastoral systems and the procedures followed can be identified in the tenets and propositions of some theories on social change and development. The purpose here is to briefly explore some views in development literature that relate to pastoral development. The extent to which attention has been paid to traditional production systems in Nigeria's pastoral and livestock policies and programmes will be discussed.

Our ultimate goal in this section is to understand the rationale for the involvement of government and change agents in altering pastoral forms of production. The impact of such interventions at both macro and micro policy levels is assessed which enables us to interpret and better understand paradoxical pastoral and livestock development interventions in Nigeria.

Modernizing Pastoral Production

The modernization approach is widely used in the formulation and execution of guided change programmes in many developing countries. By and large this perspective stresses the combining of industrialization and the use of inanimate power sources on a large scale (Bernstein, 1971, 1979; Levy, 1967; Moore, 1963).

Within the perspective of modernization is the "diffusion model" which places heavy emphasis on the direct transfer of agricultural technology from advanced countries of the west to the Less Developed Countries (LDCs) of the third world.

The diffusion of animal production techniques from the industrialised capitalist country to Nigeria would, for instance, be the way to develop the livestock industry, proponents of the diffusion model and modernization perspective would argue. Stated differently, agricultural development can be brought about through more effective dissemination of technical knowledge and a narrowing down of the differences in productivity among farmers and among regions (Ruttan, 1984; Ruttan and Hayami, 1984).

Like the diffusion model, the modernization approach has been a shining beacon in the Nigerian government's livestock and pastoral development programmes. Both have predominated as guiding principle for scientific inquiry in the agricultural sciences.

Nigerian livestock policies are usually couched in the rhetoric of modernization. Concern for the need to "modernize" the agricultural sector, particularly the livestock sub-sector, is often expressed. Attempts to modernise crop and animal production have frequently been based on the application of western technologies which are often assumed to be linked to increased and improved levels of productivity.

Programmes such as cross-breeding (both by natural and artificial means), ranching schemes, grazing reserves and feedlot operations are among the alternative strategies proposed to raise the productivity of local livestock species through the application of methods that should lead to the "modernization" of the pastoral sub-sector. (See, for example, Alao, 1975; Awogbade, 1981; and Oyenuga, 1973.)

Theoretically, the use of foreign technologies should transform pastoral production techniques from simple and "traditional" to ones based on modern and scientific methods. Moreover, the application of western technologies to pastoral production has been associated with an evolution of the Nigerian livestock production system from subsistence levels to commercial production.

More recent evidence of modernization-informed policies and programmes can be found in a position paper by David-West, then the Director of the Federal Livestock Department. He argued that the success of the Nigerian Government's effort to enhance the livestock industry lay in the "ability and acceptability of producers to practice innovative technology" (David-West, 1982:46). He, therefore, advocated making intensified efforts in technological advancement in the areas of breeding, pasture utilization and nutrition, not all of which have typically been practiced by livestock producers in Nigeria.

Policies that emphasized increased and efficient livestock production, through sedentarization, ranching and irrigation schemes (see, for example, Wallace, 1978) are conceptualized in modernization terms, as is the call for change from the present local systems of production to those based on foreign techniques.

It should be mentioned, however, that the foregoing tenets of modernization have a number of problems: pre-industrial societies do not all have identical social, economic, political and cultural structures, as is assumed by modernization and by users of this perspective. In fact, there is no reason to think that they are identical nor to assume that the forces of change are the same in different and divergent pre-industrial societies. For a detailed critique of the assumptions of modernization, see, for example, Appelbaum (1970), Bernstein (1971), Frank (1967), Roxborough (1979) and Taylor (1979).

Despite the serious criticism that continues to be levelled

against this perspective, modernizationists have not given up their arguments. Instead, they responded to their critics by devising "numerous articulations and *ad hoc* modifications of their theory in order to eliminate any apparent conflict" (Kuhn, 1970:78). Instances of this sort of response can be found in the works of Huntington and Nelson, (1976) and Almond, (1974).

Moreover, modernizationists have been able to draw up concrete plans for programme implementation that can be used to address real problems. Although the course of action suggested by modernizationists may not always be the most appropriate, it does attempt to grapple with tangible issues. It comes as no surprise, therefore, that the modernization perspective still informs many projects and agricultural development programmes in different parts of the third world. The situation of Nigerian livestock production is no exception, especially where development programmes and projects are sponsored by western-based international institutions such as the World Bank (IBRD), the International Monetary Fund (IMF) or other financial agencies.

Attempts to "modernize" livestock production may be seen as attempts to create conditions under which capitalist development could thrive. The Nigerian state, in an attempt to raise the efficiency of livestock production, does not simply embark upon policies that are stimulated by the realities that exist within Nigerian society. The interests of the Nigerian state are also influenced by what goes on outside its geographical confines.

In promoting measures that would "improve" the conditions under which livestock production takes place, the Nigerian Government may find it has embarked on a path leading towards peripheral capitalism. This develops through the development of commodity relations with the core or centre of the capitalist world.

A policy of modernizing livestock production, along with the accompanying capital-intensive projects, makes it possible for the Nigerian state to play a central role in reshaping pastoral production. The state, through its policies, acts to "promote extension and intensification of commodity production both in its own interests and in conditions where it might not be immediately profitable for private capital to invest" (Bernstein, 1979:433).

Often, domestic funds for investment programmes in improved

livestock production are in short supply. So, too, is the technical and managerial expertise needed to run these programmes. These shortages partly explain the role and interests of the core of the capitalist world in the promotion of schemes that require large inputs of capital.

> These schemes result from what Bernstein called an alliance between the state which organizes the political, ideological and administrative conditions of this form of capital penetration in peasant agriculture and the provision of technical and financial means of this penetration by either private capitals or the particular forms of finance capital represented by the World Bank and other aid agencies (Bernstein, 1979:434).

Projects such as the grazing reserve and accelerated livestock production programmes and small-holder fattening schemes are cases of the alliance between the Nigerian state and the core of the capitalist world.

The modernization of existing systems of pastoral production often call for the use of technologies that are not available locally. Such technologies are provided by corporations and international agencies usually based in the core. One might say that the technological interests pursued by international business organizations are met by the interests of the Nigerian state in the modernization of livestock production. The association that emerges from these common interests often lead to unequal and dependent relations.

The alliance between the Nigerian state and external organizations tends to generate conditions of dependency and underdevelopment (Baran, 1973; Cardoso, 1972; Dos Santos, 1970). The process of underdevelopment is facilitated by the economic role of the state.

The Nigerian state, in collaboration with agencies such as the World Bank, has tended to promote and stress the production of commodities. This emphasis tends to deepen the resulting dependency relationship. It can be exemplified by the emphasis put on cash crop production—such as palm produce, cocoa, groundnuts and hides and skins—especially in the 1950s and 1960s.

The emphasis changed, however in the 1970s when the concern for food production for national self-reliance and self-sufficiency emerged. This was accompanied by a call for the provision of production inputs and credit to the primary producers. For these concerns to be dealt with effectively, there was a need to seek support from outside. Donor agencies responded with aid programmes to support grazing reserves, Integrated Agricultural Development Projects, farm service centres and other farm improvement projects and, more recently, the Structural Adjustment loan scheme.

The reason for this emphasis on food production may be summed up by what Bernstein described as

> the chronic state of food production in many African countries, particularly the commodity production of food staples; the political instability associated with food shortages and inflationary food prices in the cities; the cost of foreign exchange for food imports to make up for shortages in domestic production (Bernstein, 1979:434).

These conditions were particularly aggravated in Nigeria in the 1970s. Then livestock production remained stagnant in relation to population increase, with high food import bills for the importation of meat and animal products as a result.

The foreign-aided agricultural projects that flooded Nigeria in the 1970s and early 1980s were intended to boost local production of both crops and livestock. As expected, they were capital-intensive and largely financed by revenues derived from crude oil production. Equipment and management personnel were brought in from abroad at prohibitive costs.

These projects were extended to many parts of the country even as food prices and food imports soared. A rapid evaluation of the impact of some of these projects indicates that much is yet to be derived from them (see Abalu and D'Silva, 1980; Nkom, 1982).

The interests of the Nigerian state in such agricultural programmes may be interpreted in terms of the theory of the development of alliances between the core and the periphery as part of the capitalist structure. The Nigerian state has been influenced by local interests of businessmen, senior civil servants and top military men to enter into certain relationships with the outside world in the field of agricultural development.

The interests of these groups in the establishment of ranches, large irrigation schemes, the importation of feeds and feed additives, among other goods and services, is reflected in the position taken by the Nigerian government in matters related to livestock. Any benefit that may arise from such alliances would not accrue to the general populace but to specific interest groups.

In general, capital invested to enhance local production of livestock has a tendency to extend and intensify commodity relations. This situation may in the long run lead to stagnation and even a decline in livestock production. The solution that the Nigerian government has been pursuing may result in the expropriation of pastoralists' livestock and their replacement by capitalist enterprises.

Perspectives on the State

Development policies shape the way economic activities are organized, especially in developing nations. The question of what and how to produce and who takes the decisions are significant social and economic problems in any economy. These questions are worked out in both the political and the production arenas. That is to say, politics are an important dimension of economic development and the state plays a central role in shaping economic growth. The various pastoral programmes are instances of the political dimension of state policies.

The role of the state in shaping and/or reshaping the activities of pastoral people is becoming increasingly crucial in determining the pattern of economic development in such societies. Although the impact of increasing state interference in the activities of agricultural producers is more evident in developing countries, a similar observation of the role of the state on pastoral communities may be made in the more advanced industrial societies (Gefu and Gilles, 1990; O'Donnell, 1980).

Another variant of externally induced intervention is the role of crisis managers assumed more recently by international financial institutions. Analytical attention has here centred on the role of the state as an "investment broker" and as an agent or "stimulator" of economic development. The role of the state in redirecting pastoralism may be regarded as promoting an interaction between the pastoral system of production and the capitalist state. The latter is viewed as "free from active control by members of the capitalist class but not free from the general interest of the capitalist class or from the structural requirement of the capitalist economy" (Koo, 1984).

The role of the state in attempting to sedentarize nomadic pastoralists in Nigeria may be viewed as an attempt to make a transition from the present pastoral system of production to that of capitalism. This transition may be accompanied by the creation of indigenous capitalist classes (O'Donnell, 1980) as a result of the state apparatus achieving "dynamic autonomy" (Trimberger, 1977). The state's new and more extensive activities may further be seen to serve capitalist needs of production within a changed economic and social context (O'Connor, 1970).

Why does the State Intervene?

Much attention has been focused on the role of the state especially on the issue of its proper role. In addition, a variety of proposed and actual interventions by international organizations and consultants have been fostered to help solve specific problems (Doornbos, 1989; Gefu and Gilles, 1990; Gedamu, 1990; Helander, 1990; Hjort af Ornäs, 1990; Mohamed Salih, 1990; Ndagala, 1990).

Insight into the livestock policies and programmes of government can be gained by examining its underlying reasoning. State intervention in the activities of pastoral producers has often been premised on the need to bring development in the form of improved production inputs and infrastructure. The ultimate goal is said to improve their well being. This expression of providing for the basic and essential needs of its citizens and its primary producers can be found in various government policy statements.

Cognizant of the growing population, coupled with the expanding industrial and agricultural sectors, the deepening crisis of pasture land, and the need for a reliable source of animal products to the population at large, the government of Nigeria has intervened in the pastoral production system to achieve higher levels of livestock production (Federal Government of Nigeria, 1981).

Among the prime objectives of the state is to raise the national supply of animal source protein available to the average citizen to meet the ever growing demand for meat and animal products. Particularly the urban, largely civil servant, population may get enhanced access to beef at lower or subsidized costs. Little benefit seems to accrue to low income populations (Horowitz, 1979).

The provision of locally produced beef may have another interest for the Nigerian state. The ultimate beneficiary could be government, which generates higher revenue (cattle tax) from increased pastoral activities as well as urban consumers . Furthermore, the conservation of foreign exchange earnings by curtailing imports of animals and animal products and the gradual attainment of self-sufficiency and self-reliance would be achieved.

The way African governments perceive livestock production problems in developing countries has been described by the UN Economic Commission for Africa (1985) as maximizing the contribution of the livestock sector to the overall socio-economic development of the country through increased supply of quality meat and milk, eliminating constraints in production and marketing and those inherent in the aggregate behaviour of producers, leading in particular to sub-optimal exploitation of range and animal resources; the improvement of efficiency especially where government funds are used.

The donor or funding agency appears to perceive the livestock problem on the continent as the need to modernize a rather backward system of raising livestock; to provide funds to carry out those actions deemed important and/or politically acceptable to the donor; and to provide technical assistance to ensure that defined actions are carried out "purely" and that aid money is spent "correctly".

In addition to economic reasons for national governments intervening in pastoral activities, is the task of the central government to integrate and, therefore, to exert some control over their pastoral populations. Because of the nature of their production systems, it could be difficult for the state to effectively control pastoralists politically and economically.

It has often been considered desirable and appropriate for central governments to adopt a strategy of facilitating the administration of pastoral communities, especially the more mobile groups. This meant that centralized control would be increased, taxes would be more effectively levied and collected, herd sizes would be better controlled within stipulated carrying capacity guidelines, while increased turn-over and commercialized production would be encouraged.

Such reasoning has often been justified by referring to the pressures and crises experienced by pastoral areas. Unprecedented population growth both of people and animals has resulted in overgrazing and degradation of the ecology. Another argument advanced maintains that pastoralists do not respond to price incentives and that it is time for people to take individual responsibility for limiting their herd sizes (Graham, 1988).

It has, therefore, frequently been suggested that the solution lies in adopting "modern" and "improved" livestock production techniques. Sedentarization of nomadic or transhumant pastoralists—often considered to be a pre-condition for livestock development—is usually promoted in order to make the job of providing essential services easier. Against this background national and international government agencies have launched their programmes of pastoral and livestock development.

If intervention is really geared toward improving the well being of pastoral producers, what then should the main ingredients of policy intervention be? This question is especially relevant as projects and programmes aimed at developing pastoral societies have failed repeatedly in many African and Asian countries. The issue can be said to cluster around what intervening agencies consider to be essential components of pastoral development.

The rise of a strong state control of the production systems of pastoralists, land tenure rights, increasing incidence of conflict between pastoral producers and settled agricultural people are among the factors limiting pastoralism today. More often than not, development planners and administrators use value judgements in deciding what is an essential pre-requisite for the improvement of the welfare of pastoralists.

Since development planning is often construed in economic terms such as increased productivity, higher turn-over, increased revenue, improved nutrition (mainly for urban dwellers), capacity utilization of range resources, economics of production, etc., little or no consideration is given to the human aspect of such endeavours.

Because of the ethnocentric tendency of development interventions, much is assumed about the felt needs and situation of project beneficiaries. At the same time, little is known about the existing framework within which production is organized by various groups of pastoralists nor is there much awareness of the way they define their strategies for meeting their production goals. Consequently, government pastoral policies seldom consider the needs and goals of the pastoralist at whom these efforts are targeted.

It thus seems that the information gap that has always existed between pastoralists' goals and the intervening agency's strategies and policy instruments will continue to affect any meaningful pastoral development effort. Without an understanding of the existing production culture of producers, no meaningful assessment can be made of how interventions being promoted by development agencies can be effective.

It is crucial that intervening agencies understand the producers' rationale and conditions under which production takes place. Planning and execution of pastoral development programmes using adequate and relevant information on the pastoralists' production environment should minimise unfavourable and unintended consequences, while enhancing the impact of the project on aggregate production and improved well being of the producers.

Policies and programmes on pastoral and livestock production have assumed that the present systems are inefficient and wasteful and need to be changed. It was believed that the changes introduced would result in increased and enhanced livestock productivity. Yet evidence abounds that these alternatives—ranching, grazing reserves, sedentarization schemes, etc.—are not better strategies. Indeed, the alternatives pursued by the Nigerian government do not result in improved local livestock production; rather they produce conditions which are at cross purposes with the goal of enhancing productivity (Ayuka, 1978; Breman and de Wit, 1983; Haaland, 1977; Helland, 1978; Hickey, 1978; Newsome, 1971).

The worsening crisis of shortages of grazing land has been summarized in a report by the Federal Livestock Department:

> By 1986 the situation (shortage of grazing land) [has] become very serious. Against 58.5 per cent of the total land areas of the country required to

> feed the national herd of 9 million head, only 50.3 per cent will actually be available for grazing. This means there will be a deficit of 8.2 per cent. Thus there will be overstocking of the rangelands, leading to heavy grazing, reduction in the protective vegetation cover over the soil, the development and acceleration of erosion and intensification of the process of desertification (Federal Livestock Department, quoted in Okaiyeto, 1982:524).

While often undermining local production systems, government action has frequently underestimated the productivity of pastoral systems. The benefits to be derived from the alternative strategies of better management and modernization have too frequently been overestimated.

The belief that pastoralism is inefficient is "resulting in mixed farming and settlement schemes that demand unnecessary and disruptive transformations in the characteristic pastoral way of life" (Hickey, 1978:95). Yet pastoral producers have been known to maximize their human and herd populations in multi-ethnic settings without causing environmental destruction or social conflict.

Modern forms of production have been developed to replace local methods of raising livestock. Policies that encourage a "western-style" system of livestock production often top the list of priorities in the government's livestock programme. These policies are clearly stated in the different National Development Plans (Federal Government of Nigeria, 1961; 1970; 1975; and 1981). The resultant programmes and projects continue to receive increased government support. Yet little evidence exists to show that modern production systems are better than pastoralists' existing strategies (Haaland, 1977).

Policy makers and researchers generally recognise the need to enhance livestock production to provide much needed animal protein. Changes are usually suggested by the Nigerian government or by research into the problems and constraints facing the livestock industry (see, for example, Gefu and Gilles, 1987).

Often interventions are made in livestock production—and other agricultural sectors—without considering locally available information and resources. The need to draw upon an existing local knowledge-base when making external interventions in an area has often been emphasized (See Ndagala, 1985; Salzman, 1985;

Swift and Maliki, 1984). Frankel made this point while assessing the Kongwa Experiment in the East African Groundnut Scheme:

> It cannot be too strongly emphasized that those who were asked to implement the plan were unable at any time to escape from the fundamental concepts on which it was originally based. Whatever mistakes were made were due primarily to the nature of the task, not to the men who had tried to carry it out (Cited in Horowitz, 1979:xii).

Similarly, the failure of the huge Mokwa agricultural project in Nigeria has been largely blamed on "a failure to improve on local African agriculture. The plain truth is that little, if anything, was known before the scheme started about the existing agriculture in the Mokwa area" (Baldwin, 1975:166). Another concrete instance of planning without knowledge!

Government intervention in pastoral production may be compared to the Mokwa and Kongwa projects. Projects aimed at enhancing the social conditions of pastoralists tend not to involve the input of the producers. More often than not development planning in general and pastoral development planning in particular is "top down," which may produce unintended effects. It has been argued that when local conditions and resources are not taken into consideration in the formulation of projects, the interests and needs of project benefactors may not be adequately served. Shear argued:

> Planning from the top down, while useful in identifying macro constraints to development, clearly has several limitations. Local societies can be changed and even destroyed by the development process. The impact which technological change brings about at the village (or to the pastoral group) is seldom well understood. And indeed the lack of local feedback and evaluation often brings about significant dislocations at the local level and can presage the doom of any large-scale development effort from the outset (Shear, 1978:4).

A similar observation has been made by Sandford concerning the incorporation of pastoralists' inputs and felt needs into plans of pastoral development:

> ... not only is the concept of fulfilling pastoralists' express needs valid in itself, but paying attention to these needs would also incorporate pastoralists' own knowledge into the planning and monitoring systems.

> There is a good deal of evidence that the usual bureaucratic procedures fail to do this and, as a consequence, planning mistakes are made and opportunities to learn lessons are missed, which need not have been if pastoralists had been properly consulted and full use made of their knowledge. However, usually in planning systems, although lip-service is paid to consulting pastoralists, there are no real incentives for planners and the pressures on them are to produce plans involving major capital expenditures quickly; less attention is paid to how effectively they can be implemented later (Sandford, 1983:260).

One of the policy objectives of the Nigerian government is to encourage "private ranching for beef, dairy, sheep and goats through the provision of improved pastures and fodder facilities for grazing, improved breeding stock and settlement schemes for the pastoral herdsmen... and to establish livestock service centres, in a bid to provide assistance to 10,000 nomadic families to settle down permanently" (Federal Government of Nigeria, 1981:106107).

During the 198185 National Development Plan period, the Federal Government planned to acquire 5 million hectares of land for implementing these programmes. Similar programmes of land acquisition were to be implemented by the state governments. Close to 1,250,541 hectares of new grazing reserves was planned while "over one million hectares were also earmarked for (re)settling the nomadic herdsmen".

This meant that one of the major problems identified was the system of livestock production practiced by the majority of Nigeria's livestock rearers. Increased production has been sought through attempts to alter the production pattern of pastoralists.

From the above, it can be observed that the Nigerian government's livestock policies have been geared toward eroding and replacing the present systems of production. Government policies toward pastoralists may be related to the notion held by technocrats that pastoral production systems are inefficienct (see David-West, 1982).

Other explanations may be found in the attempt of the state to exercise political control over these producers. To influence pastoral activities in the direction it desires the government may impose certain sanctions, such as stringent quarantine regulations, animal movement restrictions, fines and increased cattle tax on producers. Intervention may also be a response to the population pressure to free more grazing lands for agro-industrial and urban expansion.

All these factors may indicate that the relationship between science and policy is a complex mix of social, political and economic dynamics with specific interest groups lobbying for certain policies. The conflicting positions of different groups may be what McCrea and Markle (1984) described as "outcomes of political, ideological and economic relations".

The above observations reveal the inconsistencies in government interventions. We seek to explain the contradictions between the intentions of the Nigerian state to increase and enhance local production and its antagonistic policies towards local producers.

Interventions by the Nigerian state have been intended to boost local livestock production. Interventionist strategies have been adopted by both the colonial and indigenous governments of Nigeria to change the direction and form of livestock production. Unfortunately these interventions have not always resulted in the enhancement of pastoral activities, but have often had unintended consequences which adversely affected the existing production systems. Changes that are inconsistent with local strategies for coping with the instable ecology, values and customs of producers have indeed been proposed (Gefu, 1990).

One type of action seeks to replicate commercial livestock production systems as they exist in the "west". Livestock production in western countries, notably North America, Australia, New Zealand, the United Kingdom, Denmark and Spain, characteristically adopts the ranching system of herding. This form of herding, which involves the use of supplementary feeds and the use of fenced grazing has been introduced in Nigeria and numerous other pastoral communities in Africa.

The ranching schemes initiated in Nigeria by both the colonial and federal governments have been observed to be "discrepant with traditional practice both in the containment of the animals and in the production of special fodder crops on lands otherwise suited for agriculture" (Horowitz, 1979:14).

Another set of interventions concern range management practices. Specific patches of land are allocated to certain pastoral individuals and/or families and groups. Participants in such projects are usually required to give up many of their age-old pastoral practices such as their migratory strategies, or their decisions on herd size and composition. In return for their participation, pastoralists would get "improved pasture... and, in some projects, planting of nutritious fodder, veterinary services, water wells, credit, rationalized slaughter, market services and such intangibles as functional literacy" (Horowitz, 1979:18).

The Nigerian Government continued to try to attract pastoral nomads to settle permanently through the provision of infrastructural amenities such as improved roads, subsidized supplementary feeds and fodder and salt licks. The building of Livestock Improvement and Breeding Centres (LIBCs) to improve, multiply and distribute animals to producers was intended to serve as a model for those pastoralists who would avail themselves of services provided by the LIBCs. All these interventions by the state had one principal motivation: that of enhancing pastoral and livestock production.

3. The Sedentary Fulbe of Udubo Grazing Reserve

The previous chapters have clearly shown that, when intervening in pastoral production, the state has been seriously affected by the need to achieve macro-economic development of Nigeria. Pastoral policies and programmes are often geared toward serving (either consciously or unconsciously) the class interests of policy makers.

What are pastoralists' main production motives, which are either in line with or at variance with the interests of the Nigerian state? To what extent are the goals of pastoralists in consonance with those of the state? The next two chapters describe pastoral development and pastoralists' specific situation.

This chapter deals with two main ethnic groups namely Fulbe and Hausa. Generally, the former group is principally engaged in pastoral production keeping herds of cattle as well as sheep and goats. The Fulbe may also cultivate some land where and when it is feasible to raise grain for domestic consumption.

The Hausa ethnic group raises crops as their primary occupation. Stock raising activities may be combined with crop husbandry. When livestock and crop production are combined, fewer animals are kept. In broad terms, therefore, the Hausa group can be said to be a predominantly crop raising population (agriculturalists) while the Fulbe are primarily animal producers (pastoralists).

Sedentary and Nomadic Pastoralists

At the time of the dry season survey (November-December) only long-term residents were found. Their length of stay varied between 7 and 40 years. Fulbe settlements are generally found on the fringe of the grazing reserve.

Although Fulbe pastoralists take up temporary residence on the reserve where animals are corralled and milked, they have their permanent residence outside the reserve where farming activities can be undertaken. The Fulbe are aware of the regulations which prohibit crop raising activities on the grazing reserve and faithfully abide by them.

Their farming activities are confined to patches of land around their homestead. Crops cultivated include millet, guinea corn, sorghum and beans (cowpea). Maize is not as popular as these crops, probably because of the low rains and its requirements for fertilizers. Cultivated land per Fulbe household ranges from 0.5 to 4.5 hectares.

Traditional implements are employed with little use being made of animal traction. This limited use may be associated with the size of the Fulbe fields which are often small enough to be worked by household labour. Hired labour is seldom used for the same reason as the non-utilization of animal traction.

Almost all the Fulbe interviewed expressed an unwillingness to migrate from their present homesteads where they have been residing for an average of 21 years. Many were born there. They have permanent structures such as mud and thatch buildings, wells, barns, etc. In most cases they have not moved their animals beyond the vicinity of the grazing reserve and their grazing orbit does not generally exceed 18-22 kilometres. Thus it can be noted that these pastoral Fulbe belong to the settled category.

A second group of pastoralists makes use of the grazing reserve on a regular basis. It consists of migrants from near and distant places such as Potiskum, Sokoto and Saminaka in the Borno, Sokoto and Kaduna States respectively. They manage considerably larger herds of cattle and small grazing animals than do those who reside on the fringe of the reserve.

None of these migrant pastoralists, who call themselves Garkayi, were found on the reserve at the time of the dry season survey. On our visits to locations on the grazing reserve where they usually put up their homesteads we observed traces of recent pastoral and human activities around abandoned and sparse make-shift dwellings. It was reliably confirmed that these pastoralists had moved southwards with their animals, as they normally do with the onset of dry season. As the rainy season approaches they return to the grazing reserve. On the basis of the above observations, these pastoralists can be identified as nomadic pastoral producers.

Some of these nomadic pastoralists, especially those that come from Saminaka (south-western Bauchi state), do not migrate far from the reserve. Many live in villages around the grazing reserve, grazing their animals on crop residues left on fields harvested during the dry season. Crop residue forms a major source of feed for these pastoralists during the drier months. As the rains approach and as land is prepared for the new season's crops, they move into the grazing reserve to avoid crop destruction as well as to graze their stock on green fodder in the reserve (see Chapter Four for more details).

The Resource Base

The dominant ecological regime of the area under study is that of a semi-arid zone situated between latitudes 11 40′ and 11 58′ N and longitude 10 32′ and 10 48′ E. It lies at an altitude of 350-400 metres on a plain gently sloping north-eastwards at about 0.5-1.0 per cent. The mean maximum monthly temperature may reach 40° C in April with a minimum of about 12° C in December. The soil is deep and sandy.

A grazing reserve was established to provide infrastructure for the livestock of pastoralists in the area. It is drained by the Dingaya River which flows in a north-easterly direction. At the peak of the rainy season (August-September), the river may rise and flood surrounding areas. These flood-plains provide grazing during the wet season. Climatic and soil conditions encourage vegetation that varies from an open to a dense savannah woodland. Especially during the wet season there is a wide range of species of grass, most of which are annuals.

The mean annual precipitation varies between 750 and 850 mm. With the low humidity evaporation rates are high. The rains come in mid to late June and last for about three months. In the north-western and south-western parts of the grazing reserve there are pockets of wet lowlands or *fadama* and short tributaries form drainage lines.

People and stock utilize the river, streams and earth dams

during the rainy season, while they rely on ponds and dug wells for their water supply in the dry season. The area was formerly infested by tsetse fly and although it is officially claimed that they have been eradicated, some pastoralists mentioned the tsetse fly as one of the constraints in their endeavours to rear livestock. The incidence of the fly seems to be largely restricted to the flood-plains of the Dingaya River where the vegetation cover is much denser than in the other parts of the reserve.

Severe deterioration of the rangeland especially around watering points and the riverine *fadama* areas was observed. Intensive use has resulted in serious degeneration of the vegetation and degradation of the soil in large portions of the reserve. The combination of intensive grazing and trampling underfoot has frequently resulted in wind and sheet—and, in some places, gully—erosion.

Nearly all our respondents—especially the pastoral Fulbe—complained bitterly about the disappearance of highly valued and desirable forage and grass species. They claimed that the heavy influx of migrant pastoralists grazed out and destroyed forage areas and grasses like *gamba* grass. It was formerly readily available on the grazing reserve but now virtually eradicated. Most of the pastoralists interviewed acknowledged that the reserve has been over-grazed and expressed a wish that some mechanism would be devised to control the influx of "outsiders" to the grazing reserve.

The Hausa group are found in larger settlements around the reserve as well as living side-by-side with their pastoral Fulbe neighbours in small communities. Communities of Fulbe pastoralists and Hausa agriculturalists are most common along the eastern fringe of the grazing reserve between Gumai and Shelin, a distance of approximately 37 kilometres.

The Hausa group (Hausawa) own and cultivate larger fields than the pastoral Fulbe and keep fewer animals, especially cattle. Draught animals are almost exclusively kept by the Hausawa, who utilize animal traction, apparently for the simple reason that they work larger fields.

It is not unusual to find large farmlands in and around the grazing reserve. On the reserve the area between Gangawa and Shelin—a stretch of about 6-7 kilometres—is being heavily cultivated. Crop fields vary between 2 and 6 hectares. It seems that this portion of the reserve has been legally leased to and cultivated by some individuals.

The intensity of farming in this part of the grazing reserve has seriously inhibited pastoral activities. Livestock owners are forced to find alternative places to pasture their animals in order to avoid destroying crops. Some pastoralists found it difficult to understand why certain individuals were allowed to raise crops on the reserve while they themselves were forbidden to do the same.

Conflicts have been reported between livestock owners and people who raise crops on the reserve. Cases of litigation are frequent as a result of alleged crop destruction by animals usually owned by pastoral Fulbe, who are often at a disadvantage. Heavy fines have been imposed as compensation for alleged damage done to crops. The Fulbe defendent is often forced to sell an animal or two in order to adhere to court rulings.

Because of the court's tendency to find the livestock owner guilty, some agriculturalists delay harvesting their crops beyond the usual harvesting period or leave harvested crops on the fields. This may be an attempt to incriminate an owner whose livestock may graze on part of such a field or on the harvested crops.

Despite this apparent conflict, the pastoral Fulbe interviewed would say that their relationship with the agricultural Hausawa is cordial, apparently for fear of being victimized. Private discussions with the security chiefs of the Fulbe, the *Sarkin Fada*, and the heads of groups of Fulbe families, the *Ardos*, clearly indicated that there is conflict and unspoken antagonism especially when it comes to matters of land utilization.

Herd Size and Patterns of Ownership

In general, the size of the herds of settled pastoralists are smaller than those of the nomads. The settled pastoralists in and around the Udubo grazing reserve are no exception. The average herd size per household is about 32 head of cattle. They are mainly comprised of the White Fulani and Red Bororo breeds, with the White Fulani predominating.

Larger herds of cattle are found in the *fadama* areas. Both the reported and observed cattle population indicate that pastoralists in the Bakin Kogi area—a *fadama* area—have larger herds of cattle but lesser flocks of small ruminants than in other areas. Herds of 80-100 cattle are not uncommon.

A total of thirteen *Ardos* (heads of groups of Fulbe families) and *Mai Anguwa* (ward heads) residing in and around the reserve were identified and interviewed specifically on the animal holdings. They were asked about the approximate number of households in their domain and the average herd size per household. Each *Ardo* and *Mai Anguwa* estimated there were about 30 household heads under them with an average of 32 cattle each.

This indicates that a total of about 12,480 cattle were to be found in the area at the time of the dry season survey during November and December. This number increases from late June to September, during the rainy season, due to the influx of pastoral producers from other parts of the country.

The above estimation of the cattle population, based on enumeration and observation, does not pretend to represent the actual carrying capacity of the area in the dry season. If more accurate estimates of the livestock population need to be obtained, we suggest that our survey be complemented by an aerial survey.

The unit of ownership is the household and the household head is responsible for taking management decisions. Only a few of the pastoralists interviewed reported herding or keeping cattle in trust for non-household members. Civil servants and businessmen are the categories of people that often entrust cattle to pastoralists.

In nine cases of cattle entrustment, only a few animals were reported to be entrusted (between 1 and 11 cows). The exact number of animals held in trust was, however, observed to be underestimated. Pastoralists who kept cattle for other people also had animals of their own. In these cases the total number of cattle held in trust exceeded those owned by the pastoralists.

The incidence of cattle entrustment may be higher than that reported during the survey. Perhaps this happened because few pastoralists, especially not the Fulbe, want to be identified as hired herdsmen with few, if any, animals of their own. Where the herding unit is inconsistent with the ownership unit, major decisions regarding herd welfare become problematic and meaningful external interventions may be rendered less effective.

Sheep and goats are usually kept along with cattle. In most cases, small grazing annimals are herded separately from cattle. Where both small ruminants and cattle are herded together preference is usually for herding sheep with cattle. It is not common to find only goats being herded with cattle.

Sheep and goat flocks range between five and 52 animals, with a mean of approximately 16 sheep and/or goats per household. The goats bred were a local brown (a strain of the Sokoto red) and the sheep were Balami.

Table 2. *Livestock density among the settled Fulbe*

390 households x 32 head of cattle	12,480
390 households x 16 small ruminants TLU (6,240÷5)	1,248
	13,728

Like cattle, small ruminants are attended to by minors in the family who take animals out for grazing and watering. Cattle are corralled and small grazing animals are tethered at night. Where there is a shortage of family labour (family size varied between 1 and 22 with a mean of 9), sheep and goats remain tethered all day. Cut forage and kitchen waste are usually fed to small grazing animals and some form of shelter, or straw shade, is provided. Sheep and goats belong to the women and children.

The major source of cattle feed is on the free range provided by the grazing reserve. At the time of the survey most of the forage areas were dried out, so animals were usually herded along the lowland areas where better pasture could be found. Herdsboys usually cut herbaceous plants and scrub for the grazing animals.

Animals are also grazed on crop residue on harvested fields around the homestead. The crop residue derived from pastoralists' farms is often inadequate to sustain animals during the dry season. Thus there is a need to find other sources of fodder. Permission is normally sought from agriculturalists to allow pastoralists' animals access to harvested crop fields.

It is not usually easy to obtain consent to graze animals on crop residue. In some instances, fees are charged by farmers for the use of crop residue on their harvested fields. The age-old symbiotic relationship between agriculturalists and pastoralists, whereby harvested farmlands are grazed for crop residue in return for the manure from animal dropping, is fast breaking down. The relationship is taking on economic dimensions rather than the accepted practice of reciprocity.

This symbiotic relationship may be changing because where pressure on land is not great and where land fertility can be maintained by fallow methods, animal droppings may not be of great value to the agriculturalist. If, however, the pressure on land is considerable animal manure becomes a valued input on the farm (Gefu, 1986).

Most pastoralists surveyed reported spending substantial amounts on feeds, including payment for grazing on crop residue around the grazing reserve. Their general concern related not to the high cost of feeds but to its scarcity especially during the critical dry months of March to May, before the rains come.

Many pastoralists reported that they usually travel long distances to neigbouring states to buy animal feeds, particularly concentrates. A few of our respondents specifically mentioned going to Kano in search of cotton seed cake. Groundnut and cowpea crop residues (harawa) are usually available in local markets.

Stock Management Systems

The traditional free range grazing practiced by most Nigerian pastoralists has continued to contract as a result of a combination of agricultural and industrial expansion as well as population dynamics. Marginal lands that used to be grazed especially by nomadic pastoralists, are increasingly being used for non-pastoral activities, thereby heightening the marginalization of pastoral producers.

The extent of the peripheralization of pastoralists is demonstrated by the newspaper reports about conflicts between pastoralists and agriculturalists, many of which involve the loss of lives. Some of the worst conflicts have been reported in the non-traditional grazing zones. The tension that seems to build up between pastoralists and their agricultural hosts can be attributed to competition for land use. Over the years the needs and activities of pastoralists and farmers in regard to land utilization have been at real or apparent variance.

One of the steps taken by the government to minimize confrontation between these two groups was to establish grazing reserves, especially in the traditional grazing areas of the country. These reserves were meant to provide the bulk of pastoralists' grazing needs. The Udubo grazing reserve, which is primarily a wet season resource, is one of the gazetted reserves.

Regretably, the establishment of grazing reserves has not reduced the incidence of conflict between pastoral producers and agriculturalists. There is a tendency for pastoralists to continue their age-old practice of free range utilization by adopting some form of mobility and crop residue grazing as their main strategy of resource exploitation.

Since the Udubo grazing reserve offers grazing facilities mainly during the rainy season, it is to be expected that pastoralists in and around the reserve will move their animals in search of forage during the drier months. During this period, settled pastoral Fulbe as well as their Hausa agro-pastoral counterparts feed their animals principally on crop residue left on their own farms and on adjoining farmlands.

Animals also browse among plants and some grass regrowth around the *fadama* areas in the lowlands. Pastoralists residing on the fringes of the reserve, especially those on the Gumai-Shalin stretch, walk an average of 13 kilometres a day while grazing and watering their animals. Supplementary feeds include cotton seed cake, groundnut cake, harvested crop residue (*harawa*) and grain husk (*dusa*).

When the sources of water dry up towards the middle of the dry season, animals are provided with water from wells dug by individual pastoralists around the homestead *ruga*. Almost every Fulbe *ruga* visited had a well nearby. Most wells provide water all year round. In nearly all cases the well water was suitable for consumption by both livestock and humans.

During the wet season, a period that coincides with a large concentration of cattle on the reserve, water is expected to be more readily available from alternative sources (rivers or streams, ponds, lakes, dug-outs, etc.). The high concentration of stock at this time of the year seems to account for the extraordinary degree of range abuse observed in many sections of the reserve.

Systems of Land Tenure

The prudent use of land connotes a good understanding of the land tenure system in any given community. The relationship between land use and land tenure is reciprocal. The ownership of land is the primary source of socio-economic and socio-political power, particularly in agricultural societies. As ownership and control of land is skewed, so are the accompanying social, economic and political privileges unevenly distributed.

The quality and quantity of land available to pastoral producers, especially the Fulbe pastoralists, has dwindled as a result of urbanization, growth of human and animal populations, and the rapid expansion of industrial, commercial and agricultural activities. In short, the increased pressure put on land, especially rural land, in the past decade or two has been such that pastoral producers have been pushed on to peripheral lands.

Land traditionally used by livestock rearers in both upland and riverine *fadama* areas is rapidly giving way to large-scale agricultural and industrial enterprises. Examples of such capital-intensive ventures include operations of the Agricultural Development Projects (ADPs), Integrated Rural Development Projects (IRDPs) River Basin Development Authorities (RBDAs) and other irrigation schemes.

This continous encroachment on grazing territories explains the pressure on the few remaining grazing areas available, both in established grazing reserves and on other rangelands. The open conflict between pastoralists and agriculturalists is other evidence of the "land-squeeze syndrome".

The over-use of grazing areas sets in motion a chain-effect of decline in range productivity, a re-awakened need to migrate on the part of pastoral producers, increased conflict between agriculturalists and pastoralists, litigations arising from crop destruction by animals, disincentives to traditional stock rearers and a decline in the production of livestock and animal products. This situation cannot be allowed to continue. An answer to this "time bomb" lies basically in the issue of land use and the land tenure system.

In discussing the land tenure and land use system for pastoralists in and around the Udubo Grazing Reserve area, it is useful to distinguish between ownership of land and operational holdings. Land use by pastoral Fulbe is principally an operational holding while land held by Hausa pastoralists can be classified as ownership holding.

By definition, an operational holding comprises all the land an individual user is able to cultivate or utilize during any given period. It includes patches of land over which use rights are held for a specific period of time. Land over which temporary use rights are held such as leasehold (*ara*), mortgage (*jingina*), rent (*aro*) or fallow land (*daji*) are included.

Most of the pastoral Fulbe cultivate fallow lands and have done so for many years. In many instances, the land they have cultivated has not been allocated by specific community members nor by community leaders. Such land has often been occupied and cultivated by their ancestors. Fallow land cultivated by the earlier Fulbe "settlers" around the Grazing Reserve passed from one generation to the other. This does not, however, prompt the Fulbe to make claims of ownership to such land.

The Fulbe pastoralist is aware that he is only passing on operational holding privileges to his heirs and makes no land ownership claims. This indicates that the Fulbe is very much aware that he does not have a legal right to the land he presently utilizes and that the community may revoke the operational holding privileges if it so desires.

Ownership holding refers to land that can be passed on to one's heirs as inheritance. This is the Hausa agriculturalist's most common type of holding. He has ownership rights and can pass on a patch of land to his heirs.

The present land tenure arrangement is essentially based on the customary system of land tenure which was enshrined in the Land Tenure Law of 1962. The administration of all land in Northern Nigeria was based on this law after 1960.

The prevailing land tenure system has hardly been altered, even after the passing of the Land Use Act in 1978. Pastoralists are still at the mercy of their host communities. The 1978 Act does not provide traditional pastoral producers with any legal rights over land. The land question must be resolved if the livestock industry is to be saved and sustained.

Perceived Needs

> Pastoral development can be conceived as a mainly social development activity, aimed at the improvement of the standards of living of pastoralists through the provision of health, education, vetinary, water and other services together with institutional building for better systems of range management. Livestock development, on the other hand, is an economic activity, based on cost recovery (Gefu, 1991:181).

Any intervention in the present system of pastoral production should take into consideration the realities within which the pastoralist operates. Intervention should be formulated to facilitate the net gain accruing to the pastoralists and to society in general, rather than serving the needs of interest groups.

Furthermore it should be recalled that pastoralists are receptive to change when such change is compatible with the realities of their pursuits.

The Wish to be Resettled

Most of the pastoralists interviewed were settled and have been residing in the area upwards of eight years. Many were born and raised in the area in which they presently reside. Asked if they would opt for an alternative settlement if their needs could be provided for in new locations on the grazing reserve, most respondents answered in the negative. They claimed that they have become so attached to the area that they would prefer to remain there rather than risk moving to an unfamiliar place. They generally feared losing the patches of land around the homestead on which they raise crops.

Continuity of farming activities as well as maintaining family ties were repeatedly mentioned as major reasons for not wanting to be relocated. Being aware of the "no-farming" regulation on the grazing reserve, most of our respondents (especially the Fulbe) prefer to remain on the limited land they cultivate rather than to give this up for an uncertain alternative. This thinking emphasises the important role crop farming is playing in the activities of pastoral producers.

Grain for household consumption comes primarily from pastoralists' farms, so the Fulbe pastoralist rarely buys grain. There are several cases where as many as 100 bags of grain (10,000 kg) were reported to have been harvested by an individual Fulbe household in one season.

Another reason for the reluctance to relocate can be associated with the permanent structures the pastoralists have erected over the years, including dwelling places, wells for water and barns to store grain. Pastoralists have become used to being near to certain markets. All these factors combined, could seriously threaten efforts to relocate pastoralists elsewhere on the reserve.

A few pastoralists at Gamawa and Taranka, however, expressed willingness to relocate. They were Fulbe pastoralists who complained about the attitude of crop farmers. Several have been involved in court cases with agriculturalists in which fines were often imposed on the pastoralists for alleged destruction of crops.

Except for a few in Gamawa, they did not indicate any preference, but wanted to move to any place where they would have legitimate access to pasture, water as well as arable land throughout the year. Where preference was expressed, two possible settlement areas were usually mentioned, Gangantilo and Awuno, both in Borno State. These areas are said to have year-round supplies of water and pasture.

The few Gamawa Fulbe pastoralists who opted for relocation represent a departure from the general trend. This leads us to the issue of whom to consider for possible settlement and adoption of the agro-pastoralist model.

Based on the information available to us, our opinion is that nomadic pastoralists, who do not have any form of land-use rights, need be considered among the principal beneficiaries of such projects. This would not only improve the productivity of nomads' herds but could reduce the incidence of open confrontation between crop and livestock producers.

There is however a need to conduct a similar survey on the attitudes of nomadic pastoralists towards broad pastoral development proposals. Such a survey should provide coverage of all categories of pastoral producers.

Grazing

A common constraint faced by pastoralists concerns animal feeds and water supplies. All respondents complained of inadequate pasture especially in the dry season. Even during the rainy season, many of the valued forages are no longer available on the grazing reserve. The species include *alaludure, bulude, daramua, gamba, garlarba, jarugai, jemerai, jile, kajale, paguri, shinle* and *sohbe*. All these types of fodder, both grasses and legumes, were said to have been on the reserve in the past, but have been virtually eradicated as a result of over-grazing.

The pastoralists recognized and valued these fodder for their high nutritional value, which they claimed improved the productivity and health of their animals. The disappearance of these plants was usually blamed on the influx of nomadic pastoralists during the rainy season.

Water

The pastoralists interviewed did not represent their water requirements as being as acute as those of fodder. Except in the Gumai area, water for livestock was not mentioned as the most crucial problem. The problem pastoralists in the Gumai area face can partly be linked to a watering pen provided by the National Livestock Production Division, NLPD. These pastoralists had become so dependent on this source of water that they found it extremely difficult to find alternative sources when the service was discountinued.

Generally, pastoralists dig their own wells. Water is not usually bought for the day to day upkeep of animal although some may be bought when animals are taken to market to be sold.

Veterinary Care

Animal health was not frequently mentioned as a major production problem. Pastoralists generally considered their herds healthy. Our observations of the conditions under which animals were being managed confirmed this. There are, however, isolated instances of disease occurring among herds of cattle as well as among flocks of small ruminants.

Among the most common cattle diseases were *raba, shimilo, harbou* and *boru*. *Raba* is a skin condition that accompanies excessive rainfall. *Shimilo* affects the eyes and skin of cattle and is accompanied by acute diarrhoea. The disease may claim an animal's life within two days of the onset of an attack.

Another disease is *harbou*, which affects an animal, usually cattle, while grazing in the bush. It is likened to an electric shock. The animal may collapse and die on the spot. *Boru* causes swelling in the mouths of cattle. The affected animal salivates abnormally and the hoof cracks as well.

Diseases common among sheep and goats include *bugau* and *gishu*. *Bugau* is more prevalent among sheep than among goats. The affected animal becomes restless and confused and may die within a few hours if not treated. *Gishu* is a goat disease, particularly common in flocks of cross-bred goats. Which breeds of goats are more prone to this disease are not known to this researcher. It is probably because of *gishu* that pastoralists do not usually keep crossbred goats and sheep. They acknowledge that especially small cross-bred grazing animals come with their own management problems, one of which relates to health.

Most pastoralists did not seek veterinary help in the day to day routine of animal management. Although there is a veterinary clinic at Udubo, drugs and veterinary supplies are grossly inadequate and the clinic can hardly provide any form of service beyond occasional advice and vaccination campaigns in the event of the outbreak of a bovine disease.

In addition, pastoralists generally expressed the fear that they would be asked to slaughter sick animals, which has happened fairly often. Most pastoralists treat disease conditions by using local herbs and concoctions. They also purchase and use modern veterinary drugs and expressed a willingness to invest even more on the purchase of drugs if they were readily available.

A few pastoralists have become so used to administering modern drugs that they sometimes travel hundreds of kilometres to procure medicines. Pastoralists could not tell us the names of the drugs they purchased, but they recognise and are able to identify such drugs. They were generally satisfied with the response whenever veterinary assistance was needed, especially during outbreaks of epidemics such as rinderpest.

Marketing

Markets for livestock and general goods are important components of pastoral activities. The market serves as an information centre for pastoralists and most of the information they need to make management decisions is available there. Information about diseases prevailing in certain areas, available drugs, pasture, water, etc. is shared during visits to the market.

Attendance at markets may or may not include the intention to dispose of animals or animal products. When a visit to a market involves the sale of any pastoral product—live animals and/or dairy products—transportation is generally on foot. Although a few motorised vehicles run between certain points during market days, most pastoralists prefer to walk to and from the markets.

Animals culled for sale are mainly bulls. Occasionally unproductive cows and heifers are also sold. On one of our visits to a cattle market at Wabu (a fairly accessible location) bulls weighing on average about 210 kg sold for between 450 and 480 *naira*.

When pastoral Fulbe offer an animal for sale it may be a management-related decision or because of the need for cash. Cattle and small grazing animals are sold through middlemen, *dilali*, who inflate the price of the animals, as they get a commission on every animal sold.

The dairy products sold consist mainly of yoghurt (*kindirimo*), butter (*mai*) and skim milk (*nono*). Fresh whole milk (*madara*) is not customarily sold but is usually consumed in the *ruga* by household members or served to visitors. The processing and marketing of dairy products is primarily the task of women with the younger ones helping to milk the cows. At the time of survey, a litre of *kindirimo* generally sold for 80 *kobo* in the market.

There is a general desire to increase milk production and marketing provided storage does not become a problem. Pastoralists' reluctance to sell *madara* can be related to their traditional techniques of preserving dairy products which minimize the risks associated with storage. In other words, the apparent refusal to sell fresh milk is a custom which arose from the necessity to preserve dairy products in forms that can be stored for sometime.

Any increase in the volume of milk production must be matched by an efficient system of distribution. There should be collection centres within walking distance. Waste must be minimized, if not eliminated, through the prompt collection and processing of the milk produced.

Social Services

Pastoral producers are members of a larger community administered by the Gamawa Local Government Authority (LGA). Like any other productive members of a community, pastoralists expect basic social infrastructure. Health clinics, schools and water supplies are among the services pastoralists complained of not getting in the Udubo grazing reserve. Pastoralists have found a partial solution to the problem of water by digging wells, but the more complicated tasks of providing health and educational infrastructures cannot be shouldered by individuals in the way the water problem has been handled.

By providing themselves with wells pastoralists and agro-pastoralists in the area have learned to reduce their dependence on water from boreholes or other sources. The general feeling of respondents was that it is the responsibility of the local authority to provide social services. The key argument was that they have contributed as much (if not more) to different development funds as any other group of producers, without getting anything in return.

Because they expected the LGA to provide certain basic infrastructure, pastoralists were reluctant to pay for water development even within the framework of the agro-pastoralist model. A few pastoralists, especially the *Ardos* and *Sarkin Fada* of the Fulbe, questioned the moral justification of asking them to pay for what ought to have been provided at the expense of the local authority. They did not seem to understand where the levies collected from them went.

Thus, it may be difficult to persuade pastoralists to pay for water development since they tend to view it as their right.

There are a few health clinics around the grazing reserve, but none is within reasonable distance of the pastoral communities. Long distances, sometimes up to 25 kilometres in the case of pastoralists in the Jar Mari/Gangawa/Taranka area, have to be walked to get to the nearest health centre. There is a need to provide health clinics within reasonable distance of pastoral communities. The health of the pastoralists is just as important as that of their herds.

Educational facilities was another area about which pastoralists indicated concern. Particularly Fulbe pastoralists in the Maga Giri area complained about the lack of educational facilities for their children. Since the only primary school in the community was closed down as a result of a school merger, children have to attend school in distant places. Most of these children have been withdrawn from school because of the distance. In Maga Giri some interest was shown in adult literacy. Pastoralists expressed a willingness to support literacy drives through community efforts. At Gamshua the community is presently involved in development and is operating a primary cum islamic school.

Attitudes toward Development Proposals

Generally pastoralists have good management ability. They employ rational strategies for utilizing the scarce resources available to them. They are also dynamic and accept change where they can be proven to be utilitarian.

Most pastoralists interviewed expressed their willingness to spend reasonable amounts of money on the upkeep of their herds. Some were prepared to double their present expenditure on animal feeds, drugs and other necessities. Many of our respondents were willing to pay for improved pastures especially if of the more familiar fodder were available.

At the time of the survey, the purchase of animal feeds was observed to be common. Crop residue (*harawa* and *dusa*) was the most readily available and most frequently purchased animal fodder. Many pastoralists reported traveling up to Potiskum (about 65 kilometres from Udubo) to buy feeds. Some Fulbe pastoralist indicated the desire to obtain credit for the purchase of more dry season feeds. They said that if such assistance became available, they would repay their debt by selling some of their animals, especially unproductive cattle and small grazing animals as well as by raising more crops where feasible.

It is reasonable to expect that any feed development effort embarked upon in the area is likely to be utilized. The mode of operation of feed and pasture development (fodder bank) needs, however, to be carefully planned.

The fodder bank concept was differently received. While pastoralists indicated interest and willingness to pay for pasture development—given credit and timely supplies—most were apprehensive about the utility of the total package, especially as they did not want to pay for the supply of water.

They expressed resentment about any arable farming on the reserve. Some pastoralists who showed interest in growing their own fodder expressed serious concern about theft, unauthorized grazing and acts of vandalism but were not convinced that fencing would solve the problem. They did not want to invest substantially in it, as fencing is expensive and may be broken down by cattle.

There is concern that not every pastoralist would benefit from development schemes. There is a tendency for fodder grown and stored by participating pastoralists to be shared with non-participating kinsmen during critical periods. The nature of pastoral social relationship precludes a participating member of a clan from stacking fodder away for his own livestock while watching his kinsmen's animals starve.

The intended effect of establishing a fodder bank would then be defeated as the remaining fodder would not be enough for the extended families of all the participating pastoralists.

4. The Garkayi and the Grazing Reserves

It was observed in Chapter Three that two major ethnic groups were found in and around the reserve, the pastoral Fulbe and the Hausa agriculturalists. During the wet season, survey attention was focused on a third group, the nomadic pastoralists.

Our dry season respondents had mentioned five distinct groups of nomadic pastoralists who use the grazing reserve during the wet season, from June to September. The most frequently mentioned nomadic groups were the M'bororo, Yabaji and Garkayi (Garkanko'en). Our wet season survey, however, revealed there were not as many nomadic clans as we had been made to understand. It can be inferred that the settled pastoralists attributed the poor condition of the range to its use or misuse by the nomads. The claim that many nomads were using the reserve was perhaps an attempt by the sedentarized groups to exonerate themselves from responsibility for the deteriorating rangeland. The findings of our wet season survey confirmed this supposition.

At this time only one nomadic clan was found in and around the reserve. It was frequently referred to by both the settled pastoralists and the agriculturalists as Yabaji, Agoi or M'bororo. The group we met on the reserve, however, identified themselves as the Garkayi clan.

Nomadic Pastoralists

The Garkayi pastoralists trace their origin to Kano. Asked whom the Yabaji and M'bororo clans were, the Garkayi consistently responded that they were usually regarded by their hosts as belonging to either the Yabaji or M'bororo clan. Both the agro-pastoralists and the agriculturalists referred to migrant cattle herders collectively as M'bororo. They are usually seen not only as a migrant but also a pagan group and are regarded as "non-believers" and "social deviants" by the majority of the sedentary groups. One of our respondents (an *Ardo*) said, "any migrant pastoralist who does not pray, does not speak Hausa and dresses and does his hair like a woman's, is an M'bororo".

It was observed that as soon as an "M'bororo" becomes a practicing Muslim (i.e. prays regularly and follows the Koran), learns to speak Hausa and conforms to socially acceptable norms (i.e. dresses differently and "appears decent") the larger society ceases to identify him as an M'bororo. He henceforth is accorded a new identity and is now called Yabaji. Apparently from the point of view of the settled population, the term M'bororo is a derogatory appellation for social deviants.

Further investigation revealed that all the "three clans" (M'bororo, Yabaji and Garkayi) traced their origin to Kano. This is an indication that they are most likely to belong to the same group. We have opted for calling them Garkayi as this is how the people identified themselves.

During the wet season both the Udubo grazing reserve and other reserves have high concentrations of animals (Map 1).

Map 1. *Major livestock holdings (May-August)*

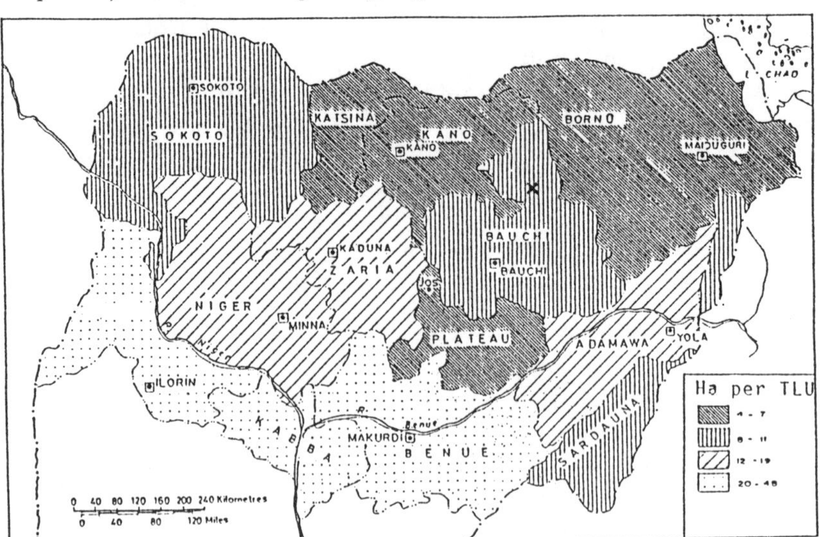

TLU = Tropical Livestock Unit = 1 head of cattle, 5 sheep or 5 goats.
X = Udubo

People of the Garkayi clan were found in different parts of the reserve but were concentrated at Taranka-Jarmari and Magajiri on the Udubo reserve and at Daji Manu, an area on a new state reserve, northwest of Udubo town. A road transverses this area from Udubo to Gadiya. These three locations formed home-bases which each group utilized extensively.

At the time of our survey, the transhumant pastoralists at Daji Manu were preparing to move to another location due to lack of water (poor rainfall) and inadequate pasture. It should be noted that whereas more plants for browsing were found at Daji Manu than on the Udubo grazing reserve there were few watering points and we saw only three earth dams there.

The nomads managed their herds in a similar manner to the system employed by their settled counterparts. After the daily milking routine both cattle and small grazing animals were taken out for free range grazing by children and young adults. Usually small ruminants were in the lead while the cattle trailed after, with the herd boys and girls following in the rear. Late in the afternoon the animals were led back to the camp, where the cattle and small grazing animals were separately corralled.

There was no form of supplementary feeding as the animals depended solely on free pasture. There were no official limitations on the extent to which the reserve could be grazed by any of the groups.

As the Garkayi pastoralists raised no crops, they depended entirely on the settled agricultural population for their cereals and other non-pastoral supplies. The women were usually actively involved in the processing and marketing of dairy products and exchanged some of these products for grain and other household dietary requirements. Because of the relatively larger herd size of the Garkayi household compared to the settled pastoralists, the marketing of dairy products was more frequent. The nomads, therefore, patronized the nearby markets more frequently to dispose of their surplus dairy products.

The extent of southerly pastoral dry season migration has considerably shifted towards the wetter and more humid areas in recent years (Maps 2 and 3). Many pastoralists are found in the rain forest zones. This may indicate that sleeping sickness or the *trypanosomiasis syndrome* is no longer a deterent to opening up non-conventional grazing areas.

During their wet season migration the Garkayi nomads we met at Magajiri and Taranka utilized the *fadama* areas around Jarmari as well as the Dingaya River banks. The main reasons for this choice that they advanced were that farming activities were minimal here, a variety of plant species was available and they did not need to fear the dreaded local legume, *zornia glochidiata* (dangere). The lowlands of the reserve had a small *zornia* population, whereas it was extensively grazed in the uplands. Water was also readily available for the use of both humans and livestock.

When the dry season sets in they move southwards to wetter areas. Those who are based in Taranka during the wet season usually move to the area round Kachia, Jere and Saminaka in the southern portion of Kaduna State for their dry season grazing.

Map 2. *Extent of southerly pastoral dry season migration (before 1970s)*

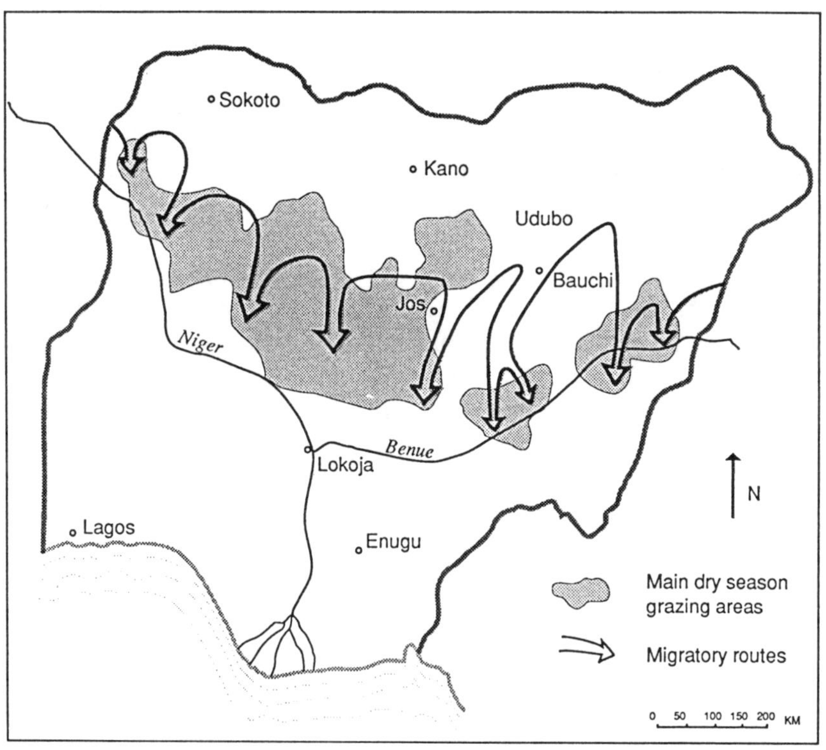

The Magajiri-based Garkayi migrate to the Kauru, Saminaka and Zaria area. It is only during the dry season that these nomads give their animals supplementary feeds, mainly crop residue which they usually buy.

The Garkayi pastoralists whom we met in and around the reserve had all the members of their household with them. No one appeared to be left hehind at their dry season base. Household size varied between 4 and 15 persons mainly children and teenagers, though a few old people were seen in some households. This observation contradicts a view expressed by other researchers who mintained that old and weak members of the pastoral nomadic groups remain behind with some cattle at their dry or wet season camps. When the Garkayi clan moves, it is usually with all household members.

Map 3. *Extent of souterly pastoral dry season migration in the 1980s.*

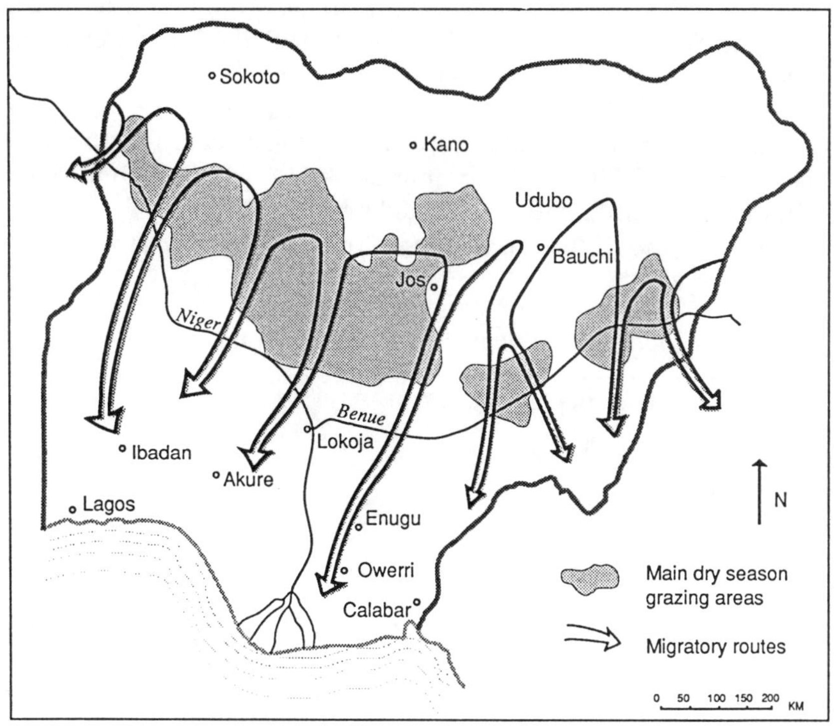

Herd Size and Patterns of Ownership

At Daji Manu, to the north of the Udubo reserve, there were an estimated 28 Garkayi households. At Taranka and Magajiri an estimate of 45 and 53 households respectively were based. They all kept cattle and sheep, mainly herds of White and Red Fulani cattle and flocks of Yankasa sheep . Goat rearing is not common, but where goats were kept the Red Sokoto (Maradi) was preferred. Cattle herds varied between 85 and 305 animals, with an average herd size of about 115 head per household. The number of small ruminants varied between 5 and 60 with a mean of about 20 animals per household. Sheep comprised about 80 per cent of this population.

The main reason for the small population of goats in the Garkayi production system was their lack of ease of mobility. The nomads considered goat keeping an additional undertaking which could impede their migration especially if they needed to relocate within a short period.

We estimated that 126 Garkayi households were staying in and around the reserve during our visit. Based on our estimate of 115 cattle and 20 small ruminants per household, there were about 14,490 cattle and 2,520 small grazing animals.

At the time of the survey, therefore, the density of stock was estimated as:

Table 3. *Livestock density among the nomadic Garkayi*

126 households x 115 head of cattle	14,490
126 households x 20 small ruminants TLU (2,520÷5)	504
	14,995

Like their settled neighbours, the nomads' unit of ownership is the household, with the head of the household being responsible for all management and marketing decisions. The entrustment of cattle was not reported among the Garkayi clan, which is understandable because of the high degree of relocation and mobility.

Perceived Needs

The Garkayi have a lukewarm interest in pursuing a settled production system. A major reason for their reluctance to settle is the deeply rooted tradition of migration which is their normal way of life. The Garkayi see themselves as cattle people and not as farmers. Traditionally they have never farmed and did not wish to start doing so now nor in the future.

The Wish to Settle

The Garkayi nomads found around Taranka and Magajiri claimed to have been utilizing the reserve during the wet season for many years. A few respondents in their late 30s and early 40s claimed they have been grazing on the reserve since childhood. Those found in Daji Manu claimed to have been using the place for over 17 years. Their temporary stay is generally between late June and late September, but varies with weather conditions. The longer the rains last and the more readily available fodder is, the longer they stay. During drier years the converse is true and they stay for a shorter period.

The attitude of the Garkayi nomads towards sedentarization can be summed up in the words of one of their *Ardos*: "Doing something different from moving our animals to pasture in free areas would amount to ceasing to exist as cattle breeders." The Garkayi were, however, quick to point out—apparently to support the wisdom of migration—that their animals are better fed, in better condition and healthier than their settled neighbours' cattle. They expressed satisfaction with their animals, which they attributed to their management system. A respected leader of the Garkayi clan mentioned that they are not "trapped" in a locality when epidemic outbreaks occur or during harsh weather conditions like drought.

Some Garkayi pastoralists, who insisted they would not settle down in any specific place even if pasture and water were made available, seemed to have little confidence that all their needs would be met if they were to take up a sedentarized life. This was apparent from their constant reference to security of life and property, "bad air", i.e. outbreaks of disease, and the lack of accessibility to land adequate for the maintenance of their present stocks.

They are able to graze and water their animals at different locations throughout the year because of the extent of their movement. The point of reference was always the condition of their animals compared to those of their settled pastoralists. They were not convinced that the settled pastoralists were doing better. Indeed, the Garkayi pastoralists were of the opinion that their settled compatriots are not true cattle breeders. They remarked on the poor condition and limited size of the herds of the settled pastoralists as well as the land squeeze they were experiencing.

The Garkayi expressed lack of trust in the Miyetti Allah Cattle Breeders Association of Nigeria (MACBAN), which was seen as an association exclusively for settled pastoralists. They did not want to be identified with an association like MACBAN. They believe it does not represent the interests of the nomads, but only those of the settled pastoralists.

When told about the intention of the government to help nomadic pastoralists to settle permanently and adopt the agro-pastoral production system, reactions were similar among clan members in all three locations. They did not want to farm at all and do not have a farming tradition, a point that they emphasized.

Furthermore, they said they did not want to be indebted to anyone because they considered debtors lazy and dependent people who are stigmatized in their society. They did not want to participate in any credit facility that may be made available. Pushed further on the issue of raising livestock in the modern way and the willingness of the government to assist, they admitted that any money given to them would be used to meet their household needs such as buying foodstuffs and clothing and purchasing more cattle (heifers) to increase their holdings. They would be able to repay the debt by selling a few bulls.

The Garkayi indicated they knew of no other wet season grazing area. Considering the poor grazing condition of the reserve, they were prepared to utilize an alternative wet season grazing if it was provided. They expressed the hope that they would be able to enjoy the things presently available to them if they were relocated. They considered the reserve adequate for their use and did not think it was terribly over-utilized but were prepared to limit their wet season grazing activities to specified areas on the reserve if required to do so.

It can be concluded that to settle these groups of pastoralists and persuade them to accept the agro-pastoralist model will require a considerable amount of hard work over a long period. The aspiration common to the Garkayi is to gradually build a large herd that would earn him recognition as a successful herder. It will take a long time exchange this attitude for acceptance of the agro-pastoralist model. Extension outreach will have to play a key role in effecting attitude change and acceptance of proven livestock production interventions.

As in the case of the sedentary pastoralists, the issue of land rights and land ownership needs to be conclusively settled if sedentarization and the agro-pastoralist model are to have any chance of succeeding. Once these basics are taken care of, other social infrastructure can be added.

Grazing

The Garkayi pastoralists utilize the reserve as a wet season grazing facility. The animals depend entirely on the natural population of annual grasses found all over the reserve. As mentioned earlier, the Garkayi pastoralists, especially those at Taranka and Magajiri, make extensive use of the vegetation in the *fadama* and along the valleys of the Dingaya River.

Here the vegetation is much denser and there are more varieties of plants than in the uplands. In the *fadama* and lowland areas, there is not much of the dreaded *zornia glochidiata* and it is mixed with other types of forage. Thus the nomads preferred to graze their cattle in these parts. Moreover, they do not have to trek long distances within the reserve since most of their animals' nutritional requirements could be met from the dense varieties of forage.

Fodder is therefore not a constraint to the nomadic pastoralists during the wet season. These pastoralists had no problem with animals becoming bloated as happened when animals grazed outside the reserve, because of the predominance of *zornia* in the upland areas, almost to the exclusion of other plant species.

The most common concern expressed by pastoralists was about shrinking corridors (cattle routes) because of the increasing encroachment of farmlands. Desperately and passionately they appealed to us to prevail on the government to stop further farming of the cattle routes. Other frequently mentioned problems during their dry season grazing were the risk of outbreaks of fire and the poor availability of fodder.

Water

Like pasture, water was not mentioned as a priority need during the wet season. The Dingaya River and its tributaries supplied much of the water needed for both humans and livestock on the reserve; while earth dams and ponds were used as watering sources by Garkayi pastoralists at Daji Manu.

Although some watering points (earth dams and ponds) may be considered unfit for human consumption, the pastoralists made do with them and were satisfied because at least their livestock could be watered. They wished, however, that a few more earth dams would be constructed so that they would not have to trek long distances from their camps in search of water.

Veterinary Care

The Garkayi pastoralists seldom mentioned animal health as a major constraint. At the nomads' camps we saw good, healthy looking animals. They combined veterinary and traditional medication. Over the years these pastoralists have come to depend on veterinary drugs and medication, especially the annual rinderpest vaccination, and on other campaigns.

The use of drugs to cope with endoparasites is becoming more popular. Some of our respondents showed us dewormers which they had purchased. There is a risk, however, of their administering expired or impotent drugs bought from local drug shops and open markets.

The most prevalent disease among their herds was *harbou,* which afflicts animals like an electric shock. Animals grazing in the bush may suddenly collapse and die within a few hours of an attack. Pastoralists believe it is due to an infliction by an evil spirit and do not know of any treatment for this condition.

Another frequently mentioned disease was *raba,* which affects the skin. This condition is treated with certain plants and herbs which the pastoralists prepare. They would not elaborate on how they made the medication locally but instead implored us to assist them to combat this skin disease with drugs.

Less than one per cent of the nomads' total stock was affected by these conditions during the rainy season. Generally the animals were in good health and looked better fed and better cared for than those of the settled pastoralists.

The Garkayi pastoralists complained of high mortality among their livestock during the dry season. One respondent lamented over the loss of 20 head of cattle at Walijo during the previous dry season. This high rate of mortality could not be attributed to any single disease. An explanatory factor for some of the deaths in the dry season may be poor nutrition.

The growing interest in and use of veterinary drugs is stimulated by the Garkayi pastoralists' contact with a missionary based at Gure near Saminaka. They talked about a "white lady at Gure who provides us with medicines and vaccines". They said that she charged minimal amounts for the drugs. Her services seemed more highly valued than those of the established government veterinary clinics. The latter, the nomads claimed, seldom offered significant help and when they did, the assistance was not as effective as that provided by the woman at Gure. They appreciated her services and wanted her to continue providing assistance.

Marketing

Due to their larger herd size the Garkayi pastoralists produce more milk and dairy products. Dairy products such as yoghurt (*kindirmu*), butter (*mai*) and skim milk (*nono*) are frequently marketed by the pastoral women. At the time of the survey, a litre of yoghurt (about 1.45 kg) sold for 60 *kobo* in rural areas and 70 *kobo* in more accessible locations. Nono sold for less, between 40 and 50 *kobo* depending on the locality.

The low price of dairy products may be related to the forces of supply and demand. While the demand seemed to remain constant, the supply was considerably inflated as both the sedentarized and transhumant herds produced more for the same markets at this time.

Social Services

Being a people on the move the Garkayi pastoralists could not be specifically considered for the kind of social services that were made available to their sedentarized compatriots. Any social services provided for the settled population may, however, be used by the nomadic pastoralists during their brief sojourn on the reserve.

One form of social service that the Federal Military Government has provided specifically for this group is the nomadic education programme. We sampled the opinion of the nomads on this programme. None of the Garkayi pastoralists interviewed knew anything about it, several months after the programme was supposed to have started.

It was generally felt that only some of their children could be released to pursue formal education. Among fairly large pastoral households the majority of the respondents split the children into schooling and non-schooling groups. The non-schooling children would be kept at home to help with the pastoral activities. Households with four or fewer children hesitated to give their consent to let their children leave home to take up formal schooling. They showed no preference as to whether a male or female child should be sent to school.

The Garkayi were asked to suggest the type of educational programme they would like government to provide for them. Where they expressed willingness to let their children have some formal education, the Garkayi insisted it must include western—

reading, writing and, most importantly, improved livestock management techniques—and Koranic components. The western education component should help them to raise their living standards and improve their earning capacity in the formal sector, while Koranic education would instil discipline and Godliness in the youth.

Education in Islamic values was considered to be a virtue which would help to reduce deliquency among their children. Cattle theft was the number one problem. Nomads thought it was made possible by the collaboration of their children with outsiders and hoped to reduce the high incidence if their children studied the Koran.

Medium and small households tended to refuse to release their children to undertake any form of education that would remove them physically from their pastoral tasks. The labour provided by children was considered more valuable than the "unseen" long-term benefits of education. For a nomadic producing group that operates under harsh conditions, it would be difficult if not impossible to find substitute labour for any children who may be sent to school.

The older folks themselves wanted Koranic and western instruction in that order. They admired Koranic scholars and the learned *mallams*, who read the Koran with ease. They, too, would like to acquire some knowledge of the Koran. They also expressed the desire to have some western education which would help them improve their pastoral production, particularly their ability to cope with problems of animal health.

The foregoing are some of the social services that could enhance pastoral activities. How this is achieved depends on the success of the nomadic education programme, which should take cognisance of the needs expressed by the intended beneficiaries especially if it is to improve their living conditions.

Other desirable social services include the provision and maintenance of cattle corridors for dry and wet season grazing. The existing cattle routes are diminishing as a result of increased farming activities. This has opened up avenues for conflict between pastoral and agricultural groups who allege that cattle belonging to pastoralists cause crop damage. Such corridors would put an end to the frequent victimization of pastoral herders by some unscrupulous mewmbers of the agricultural population.

Along with cattle routes, other livestock services could be provided to meet the needs of nomadic pastoralists. Supplementary feeds, minerals, water and medication could be made available at subsidized prices.

5. Consultation not Imposition

The livestock and pastoral policies of the colonial and indigenous administrations in Nigeria can be interpreted in terms of the theory of the complex relationship between the core and the periphery. Theoretically livestock production is re-organized to benefit the centres of the core and the periphery. The majority of measures taken by the various administrations were usually said to be directed at protecting the ecology from serious degradation. Sedentarization was thought to be the answer to resource depletion and ecological degradation.

The overgrazing and environmental depletion allegedly caused by nomadic pastoralists does not appear to be as serious as that caused by sedentarized pastoralists on the grazing reserves. Frequently warnings have been sounded about the dangers posed to the environment by migrating pastoralists (see, for example, Alao, 1975; Federal Government of Nigeria, 1981; Federal Ministry of Information, 1962; Oyenuga, 1973; Sullivan, et al., 1976). Yet the ecological degradation being caused by the activities of settled pastoral and crop producers has been ignored. The felling of trees for fuel by sedentary populations is another source of ecological degradation. It seems incredible that

> people practicing so "self-destructive" an economy could have survived. Yet they were wedded to behaviour hundreds if not thousands of years old, hewing to ancestral traditions which are, to the sympathetic but objective observer, so dysfunctional to themselves, to their habitat and to the national states in which they are subjects (Horowitz, 1979:24).

If the ecology had been impoverished over the past three decades to the extent that has been portrayed, little would be left today.

It should be emphasized that pastoralists would not like to jeopardize their livelihood by making unwise decisions regarding the utilization of the gifts of nature. They are aware of the seriousness of any act of mismanagement of resources since they are cognisant of the importance of natural grazing to their continued existence.

In planning and implementing livestock production projects, policy formulators have consistently ignored the indigenous mode of life of pastoral producers. They assume that their system is inefficient, wasteful and unproductive. In part these assumptions and negative attitudes are the result of operators of the state apparatus possessing a different socio-cultural, political and economic orientation than the pastoralists. It follows, therefore, that programmes of pastoral and livestock improvement seldom succeed and often lead to unintended and unforeseen consequences.

It should, however, be noted that state intervention is not necessarily bad nor undesirable. In as much as some intervention is necessary and probably inevitable, its adverse consequences are of great importance and what we are concerned with here.

Noting the negative effects of government intervention on pastoral production, Bourgeot observed that

> ... it is evident that this whole state of affairs, (development of capital accumulation) founded on the imperatives of capital production, is destructive of an already precarious equilibrium which has always existed, despite periodic difficulties, within traditional pastoral production—an equilibrium between the objectives of consumption and the possibilities of production. The implementation of vertical stratification, together with more widespread commercialization not under the direct control of immediate producers, lays the foundation for an "industrialized" form of pastoral production, to the detriment of its subsistence form (Bourgeot, 1981:124).

Irrespective of the intention, programmes and projects embarked upon by the Nigerian state seldom serve the immediate needs of the producers. Rather, the conditions set by donor agencies regarding admininstration of the programme are usually followed. Conditions for programmes and projects that receive agency support are often set in ways that benefit special interests in the long run.

Reviewing programmes of pastoral development in sub-Saharan Africa, Goldschmidt (1981), draws the conclusion that the schemes seldom succeeded because they attempted to impose a foreign method of production. A similar observation regarding such schemes and programmes has been made in Nigeria.

Awogbade (1981:329) noted that "programmes that are aimed at improving the quality of pastoral practices almost invariably fail". Indeed, Galaty (1981) contends that negative effects are often experienced by pastoralists and small-scale livestock producers, whose resources are diminished for the sake of commercial production.

Galaty and Goldschmidt are not alone in highlighting the undesirable consequences of sedentarizing nomads. Aronson (1980), Konczacki (1978) and Salzman (1980) have all observed that sedentarization actually inhibits efficient continuation of pastoralism and increases the degradation of the environment. This contradicts the government's intention to develop the livestock industry.

In Nigeria evidence abounds regarding the effects of sedentarization. Where sedentarization—usually spontaneous—has occurred, pastoralists are converted into agro-pastoralists with marginal outputs of both livestock and crop production. They attempt to derive the most they can from the few resources (land, water, etc.) that are available to them. They produce neither crops nor livestock in any appreciable quantity. Their reasons for taking up permanent residence differ from their desire to increase production. More often, agriculturalists turn into pastoral producers, thereby increasing the capitalization of pastoral production.

The Nigerian government has been establishing grazing reserves to provide forage for pastoralists during the drier months of the year. Serious problems of overstocking (stock pressures) and range deterioration have, however, been encountered. This is probably one setback experienced where grazing land is held in common (Artz, 1986; Gilles and Jamtgaard, 1982).

When nomadic pastoralists settle and practice crop farming, especially in semi-arid and arid zones, the combined activities of raising livestock and cultivating crops tend to increase environmental degradation. Both can certainly be profitably undertaken in less fragile ecological zones such as the southern limits of the savannah, where more vegetation is available. The major limiting factor here is the prevalence of animal disease.

The government is providing pastoralists with high-cost imported animal feeds—at subsidized rates—in an effort to induce nomads to settle or at least to minimize migration of animal and human populations. The high costs do not seem be justified. Such subsidized—and sometimes free—supplementary feeds may create permanent dependence on unsustainable subsidies. This may divert the energies of the producers from self-generating and self-reliant productive activities, by making them dependent upon external support.

Konczacki (1978) notes that the transition to semi-nomadic or sedentarized patterns of land use, by limiting movement and promoting concentration around watering points, led to the destruction of rangelands. This view has been strongly expressed by Dasman, *et al.* (1974). Overstocking and consequently overgrazing usually result in a decline of the quality of livestock following sedentarization. Sedentarization policies have often resulted in the

> abandonment of nomadic way of life which increased the incidence of coronary disease. The Somali nomads' daily intake is up to 6,200 calories and in certain parts of the country camel milk is their main diet. Yet their blood cholesterol level is low and heart disease is extremely rare. Similar findings have been made with regard to the Negev Bedouin (Konczacki, 1978:63).

The severe overgrazing evident on most of the established grazing reserves across Nigeria is a concrete result of the sedentarization programme in pastoral communitites. The deterioration of the vegetation cover to a dust bowl cannot be said to improve livestock production. An intervention that has such effects is the antithesis of ecology protection. Similarly, as there is a build-up of disease and parasitic infections around settlement areas and as a result of extremely high stock keeping rates, the health of livestock has been adversely affected. Countries like Egypt and Iran with sedentarization programmes similar to Nigeria's have likewise suffered a significant reduction in the production of meat and other livestock products.

Sedentarized pastoralists in Nigeria keep smaller herds since settling. A change from pastoralism to intensive crop and livestock raising in an area with the unstable ecology of the Sahel region could quickly turn such areas into arid lands with little or no vegetation cover. The overall impact of sedentarization may be unpleasant for pastoralists in particular and the national economy in general.

Observing the Iranian situation, Ghashgai (1981) argued that nomads have suffered economically, politically, socially and psychologically because of settlement. Sedentarization of nomadic pastoralists has been unsuccessful. There is a need to re-establish a

> proper balance between carrying capacities and actual livestock numbers within the framework of institutions that do not overlook the merits of the "traditional" and well-established systems of regulating communal grazing but take into consideration the requirements and technical possibilities offered by "modern" husbandry systems (Hrabovszky, 1981:13).

In addition, large herds are an inefficient means of exploiting rangeland vegetation. Large concentrations of animals may destroy vegetation and utilize pastures inefficiently. Cunnison notes that:

> a very large herd becomes unwieldly: the tail end struggles out of sight through the trees; towards the end of the dry season when grazing may be scarce, a large grazing herd is bad because the first cattle trample over the small patches of good grazing before the slower cattle arrive (Cunnison, 1966:68-69).

The central problem with programmes of sedentarization seems to be that of rangeland deterioration; sedentarization may, therefore, not be workable as a solution to the problems of livestock production in Nigeria.

It should be observed that commercial agricultural production does not seem to have improved the living standards of producers, but conversely has been instrumental in their impoverishment. The profits from the expansion of commercial production activities have served only to aggravate the life conditions of the producers (Bourgeot, 1981). Swift argued that increase in marketed livestock

> has not reflected any sustained increase in pastoral production; rather "the demand-led boom" in marketed livestock has been created by superimposing a modern market operation on a largely traditional pro-

duction system; this has induced a shift from a principally subsistence economy to a much more market-oriented one without real development (Swift, 1979:453).

Cole observed that Bedouin pastoralism, despite the high degree of sedentarization and specialization, is one that "does not lead to increased production of anything beyond the bare necessities" (Cole, 1981:130). This suggests that sedentarization will not necessarily result in increased holdings of livestock, as the pastoralist may not be willing to produce in excess of what he thinks can meet his needs.

From our examination of the intervention of the Nigerian state in the production activities of pastoralists and the main driving forces of pastoral production strategies, it becomes clear that the production goals of pastoralists are at variance with the intentions of external interventions. Policies and programmes aimed at changing and/or enhancing prevailing systems of pastoral production need to be conceived in the context of such systems. Distinctions made between production systems will require different sets of policies and programmes to bring about meaningful and lasting changes.

The foregoing observations show that there are many dimensions to the pastoral problem. This multi-dimensionality calls for a good grasp of the key elements that influence pastoral activities. The present study revealed that the needs of producers to meet both subsistence and non-subsistence ends constitutes the core of pastoral activities. Investigation of pastoral and livestock problems have too often been far removed from the circumstances surrounding pastoral production.

Concerns for increased livestock production have not been matched by practical efforts that could ameliorate the problems. Intervention has frequently concentrated on the use and adoption of technologies that are inappropriate to local conditions of production. It is imperative to give more attention to the human aspects of pastoral production, rather than as at present to emphasize technological policies.

Furthermore, before deciding to adopt certain technologies, careful evaluation should be undertaken to establish whether they have enhanced animal productivity and what their potential contribution might be. More importantly, technologies should be evaluated in relation to the goals and priorities of the producers.

Programmes of pastoral and rural development, aimed at improving the quality of life of pastoral producers, could be enhanced if less emphasis were given to criteria based on premises that are alien to both the producers and their environment. More utilitarian ends would be met if livestock development programmes were based on the existing pool of knowledge of the very people that these programmes and projects purport to serve.

The use of pastoralists' knowledge in formulating development programmes should be incorporated into plan protocols. Intervention programmes should reflect the interests and intentions of the producers, including their use of livestock, an important dimension in their social relationships. In other words, programmes of development should understand processes of change and the needs of the people from within, rather than imposing criteria and values from outside.

Programmes for the improvement of pastoral production should be directed toward integrating aspects of the present systems of production and making the improvements compatible with the present systems.

If state intervention in pastoral production is to benefit both the producers as well as the public at large, there must be an attempt to strike a compromise between national goals and needs and those of the producers. In so doing, an initial step would be to adequately understand all the important motives underlying the pastoral forms of production. For instance, the patterned movement of nomadic pastoralists is not migration for its own sake, but occurs because they have to move.

Since most pastoralists do not have any land rights, the only option open to them at the present time is to continue to utilize open pastures for as long as these are available. If the government feels strongly about promoting changes like the establishment of grazing reserves and sedentarization schemes, however, adequate plans and permanent arrangements must be made to facilitate all the requirements of human existence in the changed situation.

The participatory research method is favoured in the study of the pastoral people of Nigeria. This procedure utilizes multiple procedures of inquiry such as participant observation, surveys of different sorts, statistical analyses, structured and unstructured interviews, documentation and so on. This multiple method approach, known as triangulation, can profitably be used to investigate pastoral production dynamics. No single method is superior and each has its own special strengths and weaknesses (Denzin, 1978). By employing divergent approaches, the multi-faceted dimensions of pastoralism would be better understood.

We are obliged to draw inferences regarding the effects of the agro-pastoralist model on the attitudes of potential project beneficiaries. One striking observation is the need to revise and rework the fodder bank idea, if its intended purposes are to be realized.

One option would be to identify all pastoralists who belong to the same social group or clan and include them in the scheme. In this way participating social groups would be able to exercise control over pasture as a unit. This arrangement is similar to group ranching as is advocated for pastoralists in some East African countries. The management and protection of pasture land becomes a group responsibility. Since every member of the group would have a stake in the venture, all efforts should be diverted to ensure maximum protection. More care is also likely to be taken in exploiting resources over which they have control.

An alternative to the above would be to modify the *modus operandi* of the fodder bank concept. Rather than selecting a handful of "progressive/innovative" pastoralists as a model—in the hope that other pastoralists would be persuaded to adopt the agro-pastoralist package), infrastructural amenities including fodder, water, veterinary and marketing facilities should be made available to all pastoralists who are willing to take advantage of such services.

The provision of infrastructure should be on a semi-commercial or commercial basis, since most pastoralists are prepared to pay for such inputs. It follows, that the services provided should be paid for by the beneficiaries. Services may, however, be subsidized in the same way as arable farming operations are subsidized.

The introduction of the agro-pastoralist model with all its attendant initial and operating costs could entail a high risk of the participating pastoralists being over-burdened by having to repay loans, unless firm arrangements for such repayments are made. The project management may not want to pursue beneficiaries to force them to repay loans. Such action could permanently dampen future pastoral and other development efforts.

The aim of pastoral development should be explicitly stated. If, for example, the purpose of the agro-pastoralist model is to commercialize pastoral production and provide beef and animal products to urban dwellers, then the model, with a few modifications, may be considered appropriate. If, on the other hand, the aim of pastoral development is to ensure pastoralists' well-being, it may be contended that the agro-pastoralist model will not meet this goal.

This conclusion is reached because pastoralists are generally reluctant to split their herds. They are not enthusiastic about raising crops on the grazing reserve as they claim this would lead to overcrowding and damage to crops.

It is evident that the attitude of the Garkayi pastoral group to livestock production is largely at variance with development proposals. They seem to cherish large herds managed under extensive conditions. One of the hopes of the advocates of livestock development is ultimately to reduce stocking density and foster agro-pastoralism. The Garkayi pastoralists are still far removed from this orientation. There is an urgent need to embark on a serious extension programme which will persuade the nomads to take up viable livestock production techniques while remaining largely sedentarized.

In the short run, the felt needs of nomads should be attended to. This includes the demarcation of cattle routes provided with important inputs for the livestock at specific locations. As the hearts of nomads are wooed by extension personnel to modern interventions, they should gradually be persuaded to take up a sedentarized system of production. They should be assisted in organizing themselves into vigilante groups to fight the frequent livestock thefts.

To persuade both nomads and agro-pastoralists to adopt interventions such as fodder establishment and its proper utilization, the pastoralists must be overwhelmingly convinced of the benefits. Demonstration plots should convey the intended message, which is important if pastoralists are to be persuaded to utilize certain technologies. The case of neglected fodder bank demonstration plots on the reserve deserve brief mention.

Six plots of 4 ha. consisting of two plots each of *lablab* and *stylo*, transplanted *gamba* and a local species of grass called *chiasuwa* were observed at the time of the survey. These fodder bank plots were not fenced nor protected in any way. Only one plot was kept clean. In the cases we observed the plant population was low. The *stylo* plots were unimpressive as they had been overgrown by weeds. With a demonstration of this sort, not even minimum impact is likely to be achieved. There is, therefore, the need to convey convincingly the intended message of the techniques demonstrated, which can only be done by setting the right example.

When looking to the government to develop appropriate and beneficial policy programmes in the future, this study has shown that there is a need to integrate all the available knowledge on the issue. These efforts should not, however, come from one segment of the community, but from all groups and interests involved in improving conditions for both sedentary and nomadic pastoralists.

Bibliography

Abalu, G.O.I., 1980. *Farm Level Surveys of West African Agriculture: A Critique of Conventional Methodology*. New York: Purdue.

Abalu, G.O.I, and B. D., d'Silva, 1980. "Socio-Economic Analysis of Existing Farming Systems and Practices in Northern Nigeria". Paper presented at the International Workshop on Socio-economic Constraints to Development of Semi-Arid Tropical Agriculture. International Crop Research Institute of Semi-Arid Tropical Agriculture (ICRISAT), Hyderabad, India.

Adegboye, R.O. *et al.*, 1978. *A Socio-Economic Study of Fulani Nomads in Kwara State, Ibadan*. Ibadan: Department of Agricultural Economics, University of Ibadan, for the Federal Livestock Department, Kaduna.

Ademosun, A.A., 1973. "A Review of Research on the Evaluation of Herbage Crops and Natural Grasslands in Nigeria". *Tropical Grasslands*, 7 (3), pp. 285-296.

Adeyemo, R., 1984. "Agricultural Research in Nigeria: Priorities and effectiveness". *Agricultural Administration*, 17, pp. 81-91.

Alao, J. Ade, 1973. "Analysis of Some Factors Limiting the Development of the Livestock Industry in Nigeria". Paper to 9th Annual Conference of the Agricultural Society of Nigeria, July 2-6. Unpublished conference paper.

Alao, J. Ade, 1971. *Community Structure and Modernization of Agriculture: An Analysis of Factors Influencing the Adoption of Farm Practices among Nigerian Farmers*. Unpublished Ph.D. Dissertation, Cornell University.

Allan, W., 1965. *The African Husbandman*. Edinburgh: Oliver and Boyd.

Almond, G. A., *et al.* (eds.), 1973. *Crisis, Choice, and Change: Historical Studies of Political Development*. Boston.

Althusser, Louis, 1970. *For Marx*. London: Penguin.

Althusser, Louis, 1971. *Lenin and Philosophy and Other Essays*. New York: Monthly Review Press.

Amin, Samir, 1972. "Underdevelopment and Dependence in Black Africa—Origins and Contemporary Forms". *Journal of Modern African Studies*, 10, pp. 503-524.

Amin, Samir, 1973. "Transitional Phases in Sub-Saharan Africa". *Monthly Review*, 25 (5), pp. 52-57.

Amin, Samir, 1974. *Accumulation on a World Scale: A Critique of the Theory of Underdevelopment*. New York: Monthly Review Press.

Amin, Samir, 1976. *Unequal Development: An Essay on the Social Formations of Peripheral Capitalism*. New York: Monthly Review Press.
Amin, Samir, 1977. *Imperialism and Unequal Development*. New York: Monthly Review Press.
Antonio, R.J., 1983. "The Origin, Development and Contemporary Status of Critical Theory". *The Sociological Quarterly*, 24, pp. 325-351.
Appelbaum, R. P., 1970. *Theories of Social Change*. Chicago: Markham Publishing Co.
Aronson, Dan, 1980. "Must Nomads Settle: Some Notes Towards Policy on the Future of Pastoralism". In P. Salzman (ed.), 1980.
Atala, T. K., 1980. *Factors Affecting Adoption of Agricultural Innovations, Usage of Sources of Information and Level of Living in Two Nigerian Villages*. Unpublished M.S. Thesis, Ames: Iowa State University.
Awogbade, M. O., 1981. "Livestock Development and Range Use in Nigeria". In Galaty, *et al.*, 1981, pp 325-333.
Awogbade, M. O., 1982. *Fulani Pastoralism: Jos Case Study*, Zaria, Nigeria: Ahmadu Bello University Press, Zaria.
Ayuka, Lucas J., 1978. "Management of rangelands in Kenya to increase beef production: Socio-economic constraints & policies". In D.H. Hyder)ed.) Proceedings of the First International Rampeland Congress, Denver: Society for Range Management. Pp. 82-86.
Ayuko, L., 1980. "Ranch Organization Structures". Summary of papers presented at an ILCA Conference in Addis Ababa, February 25-29. *ILCA Bulletin* 9.
Bailey, Kenneth D., 1982. *Methods of Social Research*, second edition. New York: Free Press.
Baldwin, K. D. S., 1957."The Niger Agricultural Project: An Experiment in African Development". Cambridge: Harvard University Press.
Baran, Paul A., 1957. *The Political Economy of Growth*. New York: Monthly Review Press.
Baran, Paul A., 1973. "On the Political Economy of Backwardness". In Wilber, C. (ed.), *The Political Economy of Development and Underdevelopment*, pp 91-102. New York: Random House.
Barker, Jonathan, 1984. "Politics and Production". In Barker, J. (ed.), *The Politics of Agriculture in Tropical Africa*. Sage Series on African Modernization and Development, Vol. 9.
Beckman, Björn, 1981. "The World Bank and the Nigerian Peasantry". Unpublished Seminar paper presented at AKUT (Working Group for the Study of Development Strategies), Stockholm, pp. 11-22,
Bernstein, Henry, 1971. "The Modernization Theory and the Sociological Study of Development". *Journal of Development Studies*, 7 (2), pp. 141-160.
Bernstein, Henry, 1979. "African Peasantries: A Theoretical Framework". *Journal of Peasant Studies*, 6 (4), pp. 421-443.

Beal, G.M., 1969. "Some Issues We Face". *Rural Sociology,* 34 (Dec), pp. 461-475.
Berg, E., 1976. *The Economic Impact of Drought and Inflation in the Sahel* (Discussion Papers, No. 51). Ann Arbor: Center for Research on Economic Development, University of Michigan.
Bourgeot, A., 1979. "Class Structure, Political Power and Territorial Organization in the Youareg Country". In *Pastoral Production and Society.* Group for the Study of the Ecology and Anthropology of Pastoral Societies. London: Cambridge University Press.
Bourgeot, A., 1981. "Nomadic Pastoral Society and the Market". In Galaty and Salzman (eds.), 1981.
Bradby, B., 1980. "The Destruction of Natural Economy". In Wolpe, Harold (ed.) *The articulation of modes of production.* London: Routledge and Kegan Paul.
Breman, H. and C. T. de Witt, 1983. "Rangeland Productivity and Exploitation in the Sahel". *Science,* 221 (4618), pp. 1341-1347.
Brett, E. A., 1973. *Colonialism and Underdevelopment in East Africa; the Politics of Economic Change.* Heinemann, London.
Bulmer, Martin, 1982. *The Uses of Social Research.* Boston: George Allen and Unwin.
Burnham, P., 1980. "Changing Agricultural and Pastoral Ecologies in the West African Savanna Region". In Harris, D.R. (ed.). *Human Ecology in Savanna Environments,* pp. 147-170. New York: Academic Press.
Busch, L., 1980. "Structure and Negotiation in the Agricultural Sciences". *Rural Sociology,* 45, pp. 26-48.
Busch, L and W.B. Lacy, 1983. *Science, Agriculture and the Politics of Research.* Boulder: Westview Press.
Cardoso, F.H., 1972. "Dependency and Development in Latin America". *New Left Review,* 74, pp. 83-95.
Carnoy, M., 1984. *The State and Political Theory.* Princeton: Princeton University Press.
Chilcote, R. A., 1980. "Theories of Dependency: The View from the Periphery". In Vogeler, I. and A. De Souza (eds.) *Dialectics of Third World Development,* pp 299-315. Neyw York: Allanheld, Osmun and Co.
Chinchilla, N.S. and J.L. Dietz, 1981. "Toward a New Understanding of Development and Underdevelopment". *Latin American Perspectives,* Issues 30 and 31, Summer and Fall, VIII (3 & 4), pp. 138-147.
Cole, D.P., 1981. "Boudouin and Change in Saudi Arabia". In Galaty and Salzman (eds.) 1981.
Comstock, D.E. and R. Fox, 1982. "Participant Research as Critical Theory". Unpublished paper presented at the World Congress of Sociology, Mexico City.
Cunnison, I., 1966. *Baggara Arabs: Power and the Lineage in a Sudanese Nomad Tribe.* Oxford: Clarendon.
Dahl, Gudrun and Anders Hjort, 1976. *Having Herds: Pastoral Herd*

Growth and Household Economy, Stockholm Studies in Social Anthropology, No. 2. Stockholm: University of Stockholm.
Dasman, R.F., J.P. Milton, and P.H. Freeman, 1974. *Ecological Principles of Economic Development*. London: John Wiley and Sons.
de Carvalho, E.C., 1974. *"'Traditional' and 'Modern' Patterns of Cattle Raising in Southwestern Angola: A Critical Evaluation of Change from Pastoralism to Ranching"*. *The Journal of Developing Areas*, 8 (January), pp. 199-226.
Delgado. C.L., 1979. *The Southern Fulani Farming System in Upper Volta: A Model for the Integration of Crop and Livestock Production in the West African Savannah*, African Rural Economy Paper, No. 20. Ann Arbor: University of Michigan.
Denzin, Norman K., 1978. *Sociological Methods: A Source Book*, second edition. New York: McGraw Hill.
Doherty, D.A., 1979. *A Preliminary Report on Group Ranching in Narok District*, Institute for Development Studies Working Paper No. 350. Nairobi, Kenya: University of Nairobi.
Doornbos, M.R., 1989. *The African State in Academic Debate*. *Dies Natalis* Address to the Institute of Social Studies on its 37th Anniversary.
Dos Santos, T., 1970. "The Structure of Dependence". In Fann, K.T. and D.C. Hodges (eds.) *Readings in U.S. Imperialism*, Boston: Porter Sargent.
Dunbar, G.S., 1970. "African Ranches Limited, 1914-1931: An Ill-Fated Stockraising Enterprise in Northern Nigeria". *Annals of the Association of American Geographers*, 60 (1), pp. 102-123.
Durkheim, Emile, 1951. *Suicide*. New York: Free Press.
Durkheim, Emile, 1964. *The Division of Labor in Society*. New York: Free Press.
Duvall, R. and J. Freeman, 1981. "The State and Dependent Capitalism". *International Studies Quarterly*, 25, pp. 99-118.
Dyson-Hudson, N., 1970. "Factors Inhibiting Changes in African Pastoral Society: The Karimojong of Northeast Uganda". In Middleton, J. (ed.) *Black Africa: Its Peoples and their Cultures Today*. New York: Macmillan.
Dyson-Hudson, R. and N. Dyson-Hudson, 1980. "Nomadic pastoralism". *Annual Review of Anthropology*, 9, pp. 15-61.
Dyson-Hudson, R. and E. A. Smith, 1970. "Human territoriality: An ecological assessment". *American Anthropologist*, 80, pp. 21-41.
Economist Intelligence Unit, 1989. *Country report: Nigeria 1988/89*. London: EIU.
Eicher, Carl K. and D.C. Baker, 1982. *Research on Agricultural Development in Sub-Saharan Africa: A Critical Survey*. MSU International Development Paper No. 1, Dept. of Agricultural Economics. East Lansing: Michigan State University.
Etzioni, A. and E. Etzioni (eds.), 1964. *Social Change*. New York: Basic Books.
Evans, Peter, 1979. *Dependent Development: The Alliance of*

Multinational, State, and Local Capital in Brazil. Princeton: Princeton University Press.
Evans, Peter B. and M. Timberlake, 1980. "Dependency, Inequality, and the Growth of the Tertiary: A Comparative Analysis of Less Developed Countries". *American Sociological Review*, 45 (August), pp. 531-552.
Famoriyo, Segun, 1978. "Food Production Policies in Nigeria". *Food Policy*, Vol. 3 (February), pp. 50-58.
Fanon, Frantz., 1967. *The Wretched of the Earth*. New York: Grove Press.
Federal Government of Nigeria, 1970. *The Second National Development Plan*.Lagos: Federal Government Priner.
Federal Government of Nigeria, 1981. *The Fourth National Development Plan, 1981-85*. Lagos: Federal Government Printer.
Federal Livestock Department, 1984. Nigerian Livestock Information Service, 1984, Annual Bulletin, Lagos.
Federal Livestock Department, 1987. *Survey Data Report*. Abuja, Nigeria: Federal Livestock Department.
Federal Ministry of Information, Lagos, 1962. *First National Development Plan 1962-68*. Lagos: Federal Government Printer.
Federal Office of Statistics, 1976. *Nigeria Trade Summary*. Lagos: Federal Government Printer.
Food and Agricultural Organization, 1976. *Perspective Study on Agricultural Development in the Sahelian Countries 1975-1990*, 2 Vols. Rome: United Nations Food and Agricultural Organization (Publication No. PS/SAH/76/ESP/1/Ed).
Frank, André Gunder, 1967. *Capitalism and Underdevelopment in Latin America*. New York: Monthly Review Press.
Frank, André Gunder, 1969. "The Development of Underdevelopment". In Frank (ed.), *Latin America: Underdevelopment or Revolution?* New York: Monthly Review Press.
Frank, André Gunder, 1974. "Dependence is Dead. Long Live Dependence and the Class Struggle: A Reply to Critics" *Latin American Perspectives*, 1 (Spring), pp. 87-106.
Frank, André Gunder, 1979. *Economic Crisis and the State in the Third World*. Development Discussion Paper No. 30. University of East Anglia.
Frantz, C., 1978. "Ecology and Social Organization among Nigerian Fulbe". In W. Weissleder (ed.), *The Nomadic Alternative*, pp. 97-118. The Hague: Mountain.
Frantz, C., 1981a. "The Open Niche, Pastoralism and Sedentarization in the Manbila Grassland of Nigeria". In Galaty, *et al.* (eds.), 1981b.
Frantz, C., 1981b. "Fulbe Continuity and Change under Five Flags Atop West Africa: Territoriality, Ethnicity, Stratification and National Integration". In Galaty and Salzman (eds.), 1981.
Frantz, C., 1982. "Settlement and Migration Among Pastoral Fulbe in Nigeria and Cameroun". In Salzman, P.C. (ed.), 1982, pp. 57-94.
Fricke, W., 1979. *Cattle Husbandry in Nigeria: A Study of its Ecological*

Conditions and Social-Geographical Differentiations. Heidelberger Geographischen Arbeiten No. 52. Heidelberg, Germany: University of Heidelberg.

Furtado, Celso, 1964. *Development and Underdevelopment.* Berkeley: University of California Press.

Furtado, Celso, 1970. *Economic Development of Latin America; a Survey from Colonial Times to Cuban Revolution.* London: Cambridge University Press.

Galaty, J.G., 1981. "Nomadic pastoralists and social change processes and perspectives". In Galaty and Salzman (eds.), 1981.

Galaty, J.G. *et al.* (eds.), 1981. *The Future of Pastoral Peoples.* Ottawa: International Development Research Centre.

Galaty, J. G. and D. R. Aronson, 1981. "Research Priorities and Pastoral Development: What is to be Done?". in Galaty, J. G. *et al.* (eds.), 1981, pp. 14-26.

Galaty, J. G. and P. C. Salzman (eds.), 1981. *Change and Development in Nomadic and Pastoral Societies.* Leiden: E. L. Brill.

Gedamu, Fecadu, 1990. "Pastoralists and the State in Ethiopia". Paper to *Workshop on Pastoralism and the State in African Arid Lands,* September 6-9, 1990. The Scandinavian Institute of African Studies and Dept. of Social Anthropology, University of Gothenburg, Sweden.

Gefu, Jerome O. 1986. *Socio-economic information of Udubo Grazing Reserve, Bauchi State, Nigeria. Grazing Resource Survey.* Draft report to National Livestock Projects Department, Kaduna.

Gefu, Jerome O., 1987. "Land Utilization Pattern of Nomadic Pastoral Producers in Northern Nigeria: The Sedentarization Issue Revisited". In Mortimore, M., et. al., (eds.), *Perspectives on Land Administration in Northern Nigeria,*pp. 72-81. Kano: Bayero University.

Gefu, Jerome O., 1990. "Ecological and Pastoral Production in Nigeria: Issues and insights". In Baxter P.T.W. (ed) *When the Grass is Gone: Development Intervention in African Arid Lands.* Seminar Proceedings No. 25. Uppsala: The Scandinavian Institute of African Studies.

Gefu, Jerome O. and I. F. Adu, 1984. "Understanding Small Ruminant Production in Northern Nigeria". *World Review of Animal Production,* 20 (3), pp. 35-38.

Gefu, Jerome O. and J.L. Gilles, 1987. "Problems of Livestock Research and Development in Nigeria". *Samaru Journal of Agricultural Education.* 1 (2), pp. 7279.

Gefu, Jerome O. and J.L. Gilles, 1990. "Pastoralists, Ranchers and the State in Nigeria and North America: A Comparative Analysis". *Nomadic Peoples* 25-27, pp. 34-50..

Ghashghai, H.B., 1981. "The Question of Settlement of Nomads of Iran". Ph.D. Dissertation Abstract. University Microfilms International, Ann Arbor, Michigan, USA.

Gilles, J.L. and K. Jamtgaard, 1982. "Overgrazing in Pastoral Areas". *Nomadic Peoples*, 10 (April), pp. 1-10.
Goldschmidt, W., 1981. "The Future of Pastoral Economic Development Programs in Africa". In Galaty, J.G. *et al.* (eds.), 1981.
Graham, O., 1988. *Enclosure of the East African Rangelands: Recent Trends and their Impact*. Overseas Development Institute, Pastoral Development Network Paper No. 25a. London: University of London.
Haaland, G., 1977. "Pastoral Systems of Production: The Sociocultural Context and Some Economic and Ecological Implications". In O'Keefe, Phil and Ben Wisner (eds.), *Land Use and Development*, pp. 179-193. London: International African Institute.
Halderman, J.M., 1985. "Problems of Pastoral Development in Eastern Africa". *Agricultural Administration*, 18, pp. 199-216.
Haru, T., 1981. *The Social Basis of Three Development Theories*. Doctoral Dissertation, Department of Sociology. Columbia: University of Missouri.
Helander, Bernhard, 1990. "Getting the Most Out of It: Nomadic Health Care Seeking and the State in Southern Somalia". (See Gedamu, above).
Helland, Johan, 1978. "Sociological Aspects of Pastoral Livestock Production in Africa". In Hyder, D. N. (ed.) *Proceedings of the First International Rangeland Congress*, August 14-18, pp.79-81. Denver, Colorado.
Hickey, J.V., 1978. "Fulani Nomadism and Herd Maximization: A Model for Government Mixed Farming and Ranching Schemes". In Hyder, D.N. (ed.) (See Helland, above).
Hill, P., 1977. *Population, Prosperity and Poverty: Rural Kano 1900 and 1970*. London: Cambridge University Press.
Hjort, Anders, 1981. "Herds, Trade and Grain: Pastoralism in A Regional Perspective". In Galaty, J.G. *et al.* (eds). 1981.
Hjort, Anders, 1986. "Pastoral Life in Drylands". Paper to SAREC Workshop on *A Research Programme on Deforestation and Desertification*. Stockholm: SAREC.
Hjort af Ornäs, Anders, 1988. "Sustainable Subsistence in Arid Lands, the Case of Camel Rearing". In Hjort af Ornäs, A. (ed.). *Camels in Development: Sustainable Production in African Drylands*, pp 31-42. Uppsala: The Scandinavian Institute of African Studies.
Hjort af Ornäs, Anders, 1989. "Environment and Security of Dryland Herders in Eastern Africa". In Hjort, A. and Mohamed Salih (eds.). *Ecology and Politics, Environmental Stress and Security in Africa*. Uppsala: The Scandinavian Institute of African Studies.
Hjort af Ornäs, Anders, 1990. "The Cow and the Crown: What can African States Learn from European and Scandinavian Mismanagement?". (See Gedamu, above).
Hobbs, D.J., 1980. "Rural development: Intentions and Consequences". *Rural Sociology*, 45 (1), pp. 7-25.

Hopen, C.E., 1958. *The Pastoral Fulbe Family in Gwandu*. London: Oxford University Press.
Horowitz, M. (ed.), 1976. *Colloquium on the Effects of Drought on the Productive Strategies of Sudano-Sahelian Herdsmen and Farmers*. Binghamton, New York: Institute for Development Anthropology.
Horowitz, M., 1979. *The Sociology of Pastoralism and African Livestock Projects*. AID Program Evaluation Discussion Paper No. 6. Washington: USAID.
Hrabovszky, J.P., 1981. "Livestock Development: Toward 2000; with Special Reference to Developing Countries". *World Animal Review*, 40, pp. 2-12.
Huntington, S. P. and J. M. Nelson, 1976. *No Easy Choice: Political Participation in Developing Countries*. Cambridge, Mass: Harvard University Press.
Johnson, D.L., 1969. *The Nature of Nomadism: A Comparative Study of Pastoral Migrations in Southwestern Asia and Northern Africa*. Chicago: University of Chicago Press.
Keay, R.W., 1953. *An Outline of Nigerian Vegetation*, second edition. Lagos: Government Printer.
Konczacki. Z.A., 1978. *The Economics of Pastoralism: A Case Study of Sub-Saharan Africa*. London: Frank Case.
Koo, Hagen, 1984. "World System, Class and State in Third World Development". *Sociological Perspectives*, 27 (1), pp. 33-52.
Krader, Lawrence, 1959. "The Ecology of Nomadic Pastoralism". *International Social Science Journal*, 11 (4), pp. 499-510.
Krippendorff, K., 1980. *Content Analysis: An Introduction to its Methodology*. Beverly Hills: Sage Publications.
Kuhn, T.S., 1970. *The Structure of Scientific Revolutions*, second edition. Chicago. University of Chicago Press.
Lacy, W.B., L. Busch, and P. Marcotts, 1983. *The Sudan Agricultural Research Corporation: Organization, Practices, and Policy Recommendations*. Department of Sociology, College of Agriculture. Lexington: University of Kentucky.
Leibenstein, Harvey, 1957. *Economic Backwardness and Economic Growth*. London: Cambridge University Press.
Levy, M.J., 1967. "Social Patterns (Structures) and Problems of Modernization". In Moore, W.E. and R.M. Cook (eds.). *Readings on Social Change*. Englewood Cliffs, New Jersey: Prentice-Hall.
Leys, Colin, 1976. "The Overdeveloped Post-Colonial State: A Reconsideration". *Review of African Political Economy*, 5, (January-April).
Lionberger, H.F. and H.C. Chang, 1970. *Farm Information for Modernizing Agriculture: The Taiwan System*. New York: Praeger.
Lipton, M., 1977. *Why Poor People Stay Poor: Urban Bias in World Development*. Cambridge, Mass: Harvard University Press.
Lovejoy, Paul E. and S. Baire, 1976. "The Desert-Side Economy of the Central Sudan". In Glantz, M. (ed.) *The Politics of National Disaster:*

The Case of the Sahel Drought, pp. 145-175. New York: Praeger.
Luxemburg, Rosa, 1951. *The Accumulation of Capital*. London: Routledge and Kegan Paul.
Luxemburg, R., 1972. "Anti-Critique". In Tarbuck, K. (ed.) *Imperialism and the Accumulation of Capital*. London: Allen Lane The Penguin Press.
McCrea, F.B. and G.E. Markle, 1984. "The Estrogen Replacement Controversy in the U.S.A. and U.K: Different Answers to the Same Question?" *Social Studies of Science*, 14 (1984), pp. 1-26.
Meillassoux, Claude, 1974. "Development or Exploitation: Is the Sahel Famine Good Business?" *Review of African Political Economy*, 1, pp. 27-33.
Ministry of Agriculture and Natural Resources, Lagos, 1973. Survey of the Agricultural Potential of the Mambila Plateau in the Context of Nigerian Economy, 2 Vols. Repr. 1976. Rome: LI, DE.CO., S.p.A. (Livestock Development Co.).
Mohamed Salih, M.A., 1990. "Pastoralism and the State in African Arid Lands: An Overview". (See Gedamu, above).
Mohamed Salih, M.A., (Forthcoming). *Pastoral Development Policy and Options in the Sudan*. Overseas Development Institute, Pastoral Development Network. London: University of London.
Mohamed Salih, M.A., (Unpublished manuscript) *Pastoral Development and Indigenous Knowledge: Gidan Magagia Grazing Reserve, Northern Nigeria.*
Mohammed, A.N., 1976. "Nomadic Herders: Notes for Discussion". In Kumo, S. and Y.A. Abubakar (eds.) *Research and Public Policy in Nigeria: The Report of a Conference on Research and Public Policy: Priorities and Strategies*. March 22-25, pp. 233-236. Institute of Administration. Zaria, Nigeria: Ahmadu Bello University.
Monod, T. (ed.), 1975. *Pastoralism in Tropical Africa*. Thirteenth International African Seminar held in Niamey, December 13-21. London: International African Institute and Oxford University Press.
Monu, E. D., 1980. "Acceptance of Appropriate Technology Among Peasant Farmers in Nigeria". Unpublished paper presented at the Fifteenth World Congress of Rural Sociology, August 7-12. Mexico City.
Moore, W.E., 1963. *Social Change*. Englewood Cliffs, New Jersey: Prentice-Hall.
Ndagala, D. K., 1985. "Local Participation in Development Decisions". *Nomadic Peoples*, 18 (June), pp. 3-6.
Ndagala, D.K., 1990. "Pastoralists and the State in Tanzania". (See Gedamu, above).
Nelson, C. (ed.), 1973. *The Desert and the Sown: Nomads in the Wider Society*. Institute of International Relations, Research Series No. 21. Berkeley: University of California.
Nestel, B. *et al.*, 1973. *Animal Production and Research in Tropical Africa*.

Report of the Task Force Commissioned by the African Livestock Sub-Committee of the Consultative Group on International Agricultural Research.

Newsome, A. E., 1971. "The Research and Practical Experience Gained in the Development of Semi-Arid Areas of Australia". *Botswana Notes and Records*, Special Edition, 1, pp. 202-218.

Nigerian Year Book, 1982. *A Record of Events and Developments*. Apapa, Nigeria: Times Press Limited.

Nkom, Steve, 1982. "Government Policy, Rural Development, and the Peasantry in Nigeria: Mobilization or Marginalization". In Odekunle, F. (ed.) *Mobilization of Human Resources for National Development*. Proceedings of A National Workshop. Zaria, Nigeria: Ahmadu Bello University.

Nolan, M.F. and J.F. Galliher, 1973. "Rural Sociological Research and Social Policy: Hard Data, Hard Times". *Rural Sociology*, 38, 4 (Winter).

Nolan, M.F., *et al.*, 1975. "Rural Sociological Research, 1966-1974: Implications for Social Policy". *Rural Sociology*, 40 (4), pp. 435-454.

Norman, D.W., 1973. *Methodology and Problems of Farm Management Investigations: Experiences from Northern Nigeria*. Department of Agricultural Economics, African Rural Employment Paper No. 8. East Lansing: Michigan State University.

O'Connor, James, 1970. "The Fiscal Crisis of the State". *Socialist Revolution*, 1 (1, 2), January-April.

Odell, M.L. and M.J. Odell Jr., 1980. *The Evolution of a Strategy for Livestock Development in the Communal Areas of Botswana*. Overseas Development Institute, Pastoral Network Paper No. 10b. London: University of London.

O'Donnell, G., 1980. "Comparative Historical Formations of the State Apparatus and Socio-Economic Change in the Third World". *International Social Science Journal*, 32, pp. 717- 729.

Offe, Claus, 1975. "Theses on the Theory of the State". *New German Critique*, 6 (Fall), pp. 137-147.

Okaiyeto, Peter O., 1982. *Cattle Production Strategies of the Pastoral Fulani in Nigeria: A Theoretical Analysis*. Unpublished paper 1982. National Aniamal Production Research Institute, Zaria, Nigeria.

Olayide, S.O., 1976. *Economic Survey of Nigeria, 1960-1975*. Ibadan: Aromolaran Publishing Company.

Olayide, S.O. and J.K. Olayemi, 1975. "Economics of Urban Dairies in Southern States of Nigeria". *Nigerian Journal of Animal Production*, 2 (2), pp. 252-263.

Ollawa, P.E., 1983. "Focus on the Political Economy of Development: A Theoretical Reconsideration of Some Unresolved Issues". *African Studies Review*, XXVII (1), pp. 125-153.

Oram, P.A. and V. Bindlish, 1981. *Resource Allocations to National Agricultural Research: Trends in the 1970s*. The Hague, Netherlands: International Service for National Agricultural Research.

Oxby, Clare, 1975. *Pastoral Nomads and Development: A Selected Annotated Bibliography with Special Reference to the Sahel.* London: International African Institute.

Oxby, Clare, 1984. "Settlement Schemes for Herders in the Subhumid Tropics of West Africa: Issues of Land Rights and Ethnicity". *Development Policy Review,* 2 (2), pp. 217-233.

Oyenuga, V.A., 1973. "Intensive Animal Production on a Subsistence Scale". In Reid, R.L. (ed.), *Proceedings of a Third World Conference on Animal Production,* pp. 393-400. Melbourne.

Parsons, Talcott, 1951. *The Social System.* Glencoe, Ill: Free Press.

Petras, James, 1976. "State Capitalism and the Third World". *Journal of Contemporary Asia,* 6.

Petras, James, 1978. *Critical Perspectives on Imperialism and Social Class in the Third World.* New York: Monthly Review Press.

Phillips, John, 1961. *The Development of Agriculture and Forestry in the Tropics: Patterns, Problems and Promise.* London: Faber and Faber.

Pineiro, M. and E. Trigo, 1981. "Dynamics of Agricultural Research Organization in Latin America". *Food Policy,* 6 (February), pp. 2-10.

Poulantzas, Nicos, 1968. *Political Power and Social Classes.* London: New Left Books.

Poulantzas, Nicos, 1969. "The Problems of the Capitalist State". *New Left Review,* 58, pp. 67-78.

Poulantzas, Nicos, 1974. "Internationalization of Capitalist Relations and the Nation-State". *Economy and Society,* 3, (2).

Pratt, D.J. and M.D. Gwynne (eds.), 1977. *Rangeland Management and Ecology in East Africa.* London: Hodder and Stoughton.

Riesman, P., 1977. *The Fulani in a Development Context: The Relevance of Cultural Traditions for Coping with Change and Crisis.* Mimeo.

Rodney, Walter, 1972. *How Europe Underdeveloped Africa.* London: Bogle L'Ouverture Publications.

Rogers, E.M. and F.F. Shoemaker, 1971. *Communication of Innovations: A Cross-Cultural Approach.* New York: Free Press.

Rostow, W.W., 1960. The Stages of Economic Growth. London: Cambridge University Press.

Roxborough, Ian, 1979. *Theories of Underdevelopment.* London: Macmillan Press.

Ruttan, V.W., 1984. "Models for Agricultural Development". In Eicher, C.K. and J.M. Staaz (eds.), *Agricultural Development in the Third World,* pp. 38-45. Michigan State University, Ann Arbor, USA.

Ruttan, V.W. and Y. Hayami, 1984. "Induced innovation model of agricultural development". In Eicher C.K. and J.M. Staaz (eds.). (See Ruttan, above.)

Salisburg, Glenn W., 1980. *Research Productivity of the State Agricultural Experiment Station System: Measured by Scientific Publication Output.* Agricultural Experiment Station, Bulletin 762. Urbana: University of Illinois.

Salzman, P.C. (ed.), 1980. *When Nomads Settle: Processes of*

Sedentarization as Adaptation and Response. New York: J.F. Bergin (Praeger Publishers).
Salzman, P.C., 1982. *Contemporary Nomadic and Pastoral Peoples: Africa and Latin America*. Studies in Third World Societies, No. 17.
Salzman, P.C., 1985. "Local Participation in Development Decisions". *Nomadic Peoples*, 18 (June), pp. 1-3.
Sandford, S., 1983. *Management of Pastoral Development in the Third World*. London: ODI/John Wiley & Sons.
Sandford, S., 1976. "Pastoralism under Pressure". *ODI Review*, 2, pp. 45-68.
Sandford, S., 1977. *The Design and Management of Pastoral Development: Human Pastoral Populations*. Overseas Development Institute, Pastoral Network Papers, No. 2c. London: University of London.
Saul, John S., 1983. "The State in Post-Colonial Tanzania". In D. Held, *et al.* (eds.). *States and Societies*. Oxford: Martin Robertson and London: Open University.
Schneider, H.K., 1962. "Pakot Resistance to Change". In Bascom, W. and M. Herskovits (eds.), *Community and Change in African Cultures*. Chicago: University of Chicago Press.
Schneider, H.K., 1981. "The Pastoral Development Problem". In Galaty, J.G. and P. Salzman (eds.) 1981.
Scott, E. P., 1979. "Land Use Change in the Harsh Lands of West Africa". *African Studies Review*, 22, pp. 1-24.
Sewell, W.H., 1965. "Rural Sociology Research 1936-1965". *Rural Sociology*, 30 (December), pp. 428-451.
Shear, David, 1978. "The role of the Village in Sahelian Development". Unpublished manuscript.
Simpson James R. and Phyb Evangelou, (eds.) 1984. *Livestock Development in Sub-Saharan Africa: Constraints, Prospects, Policy*. Boulder, USA: Westview Press.
Smelser, Neil J., 1963. *Theory of Collective Behavior*. New York: Free Press.
Smelser, N.J., 1966. "The Modernization of Social Relations". In Weiner, M. (ed.) *Modernization*. New York: Basic Books.
Spencer, Herbert, 1958. *First Principles of a New System of Philosophy*. New York: De Witt Revolving Fund.
Spencer, P., 1974. "Drought and the Commitment to Growth". *African Affairs*, 73, pp. 419-427.
Spooner, B., 1973. *The Cultural Ecology of Pastoral Nomads*. Modules in Anthropology, No. 45. Reading, Mass: Addison-Wesley.
Stenning, D.J., 1959. *Savannah Nomads. A Study of the Wodaabe Pastoral Fulani of Western Bornu Province*. London: Oxford University Press.
Sule, B. *et al.*, 1976. "Nomadism and the Nomadic Pastoralists of Nigeria". In Kumo, S. and Y.A. Abubakar (eds.) pp. 237-247. (See Mohammed A.N. 1976, above.)
Sullivan, G.M., M.J. Yetley and W.J. Njukia, 1978. "A Socio-economic Analysis of Technology Adoption in an African Livestock Industry".

Quarterly Journal of International Agriculture, 17, pp. 150-160.
Sullivan, G.M., *et al.*, 1980. "Transforming a Traditional Forage/Livestock System to Improve Human Nutrition in Tropical Africa". *Journal of Range Management*, 33 (3), pp. 174-178.
Swift, Jeremy, 1979. "The Development of Livestock Trading in a Nomadic Pastoral Economy: The Somali Case". In *Pastoral production and society: Proceedings of the International Meeting on Nomadic Pastoralism*. London: Cambridge University Press.
Swift, Jeremy, 1972. "Disaster and a Sahelian Nomad Economy". In David Dalby and R. J. Harrison Church, *Report on the 1973 Symposium on Drought in Africa*, pp. 71-78. Centre for African Studies. London: University of London.
Swift, J. and A. Maliki, 1984. *A Cooperative Development Experiment Among Nomadic Herders in Niger.* Overseas Development Institute, Pastoral Development Network Paper No. 18c. London: University of London..
Taylor, John, 1979. *From Modernization to Modes of Production: A Critique of the Sociologies of Development and Underdevelopment*. London: Macmillan.
Trimberger, K.E., 1977. "State Power and Modes of Production: Implications of the Japanese Transition to Capitalism". *Insurgent Sociologist*, 7, pp. 85-98.
United Nations Economic Commission for Africa, 1985. *Comprehensive Policies and Programmes for Livestock Development in Africa: Problems, Constraints and Necessary Future Action*. Network Paper No. 5. International Livestock Centre for Africa (ILCA), Addis Ababa, Ethiopia. African Livestock Policy Analysis Network (ALPAN).
UNESCO/UNEP/FAO, 1979. "Tropical Grazing Land Ecosystems: A State-of-Knowledge Report Natural Research", No. XVI. UNESCO, Paris.
United States Department of Commerce, 1983. *World Population: Recent Demographic Estimates for the Countries and Regions of the World*, p.133. Washington: Bureau of Census.
van Raay, J.G.T., 1973. *Rural Planning in a Savannah Region*. Rotterdam, Holland: University Press.
Voh, J.P., 1979. An Exploratory Study of Factors Associated with Adoption of Recommended Farm Practices Among Giwa Farmers. Institute for Agricultural Research, Samaru Miscellaneous Paper No. 73. Zaria, Nigeria: Ahmadu Bello University.
von Kaufman, R., J. In Heyer, J.K. Maitha and W.M. Senga (eds.), 1976. *Agricultural Development in Kenya*. Nairobi: Oxford University Press.
Wall, 1988.
Wallace, Tina, 1981. "The Kano River Project, Nigeria: The Impact of an Irrigation Scheme on Productivity and Welfare". In Heyer *et al. Rural Development in Tropical Africa*, pp. 281-305. New York: St Martin's Press.

Wallerstein, Immanuel, 1970. *The Capitalist World Economy.* London: Cambridge University Press.
Waters-Beyer, Ann, 1988. Dairying by Settled Fulani Agropastoralists in Central Nigeria. Kiel: Wissenschaftsverlag Vauk.
Watts, Michael, 1983. *Silent Violence, Food, Famine and Peasantry in Northern Nigeria.*Berkley: California University Press.
Winrock International, 1978. *The Role of Ruminants in Support of Man.* Morrilton, Arkansas: Petit Jean Mountain.
World Bank, 1981. Agricultural Research: Sector Policy Paper. Washington, D.C.
World Bank, 1989. Sub-Saharan Africa: From Crisis to Sustainable Growth. A Long-Term Perspective Study, p.218. Washington D.C.

Personal Communications

Artz, Neal, personal communications, Colombia, Missouri, USA, 1986.

Research Reports available for purchase from the Institute

10. Linné, Olga, *An Evaluation of Kenya Science Teacher's College*. 67 pp. Uppsala 1971. SEK 45,-.
11. Nellis, John R., *Who Pays Tax in Kenya?* 22 pp. Uppsala 1972. SEK 45,-.
13. Hall, Budd L., *Wakati Wa Furaha. An Evaluation of a Radio Study Group Campaign*. 47 pp. Uppsala 1973. SEK 45,-.
14. Ståhl, Michael, *Contradictions in Agricultural Development. A Study of Three Minimum Package Projects in Southern Ethiopia*. 65 pp. Uppsala 1973. SEK 45,-.
15. Linné, Olga, *An Evaluation of Kenya Science Teacher's College*. Phase II 1970-71. 91 pp. Uppsala 1973. SEK 45,-.
21. Ndongko,Wilfred A., *Regional Economic Planning in Cameroon*. 21 pp. Uppsala 1974. SEK 45,-. ISBN 91-7106-073-1.
22. Pipping-van Hulten, Ida, *An Episode of Colonial History: The German Press in Tanzania 1901-1914*. 47 pp. Uppsala 1974. SEK 45,-. ISBN 91-7106-077-4.
23. Magnusson, Åke, *Swedish Investments in South Africa*. 57 pp. Uppsala 1974. SEK 45,-. ISBN 91-7106-078-2.
24. Nellis, John R., *The Ethnic Composition of Leading Kenyan Government Positions*. 26 pp. Uppsala 1974. SEK 45,-. ISBN 91-7106-079-0.
25. Francke, Anita, *Kibaha Farmers' Training Centre. Impact Study 1965-1968*. 106 pp. Uppsala 1974. SEK 45,-. ISBN 91-7106-081-2.
26. Aasland, Tertit, *On the move-to-the-Left in Uganda 1969-1971*. 71 pp. Uppsala 1974. SEK 45,-. ISBN 91-7106-083-9.
27. Kirk-Greene, A.H.M., *The Genesis of the Nigerian Civil War and the Theory of Fear*. 32 pp. Uppsala 1975. SEK 45,-. ISBN 91-7106-085-5.
28. Okereke, Okoro, *Agrarian Development Programmes of African Countries*. 20 pp. Uppsala 1975. SEK 45,-. ISBN 91-7106-086-3.
29. Kjekshus, Helge, *The Elected Elite. A Socio-Economic Profile of Candidates in Tanzania's Parliamentary Election, 1970*. 40 pp. Uppsala 1975. SEK 45,-. ISBN 91-7106-087-1.
37. Carlsson, Jerker, *Transnational Companies in Liberia. The Role of Transnational Companies in the Economic Development of Liberia*. 51 pp. Uppsala 1977. SEK 45,-. ISBN 91-7106-107-X.
40. Ståhl, Michael, *New Seeds in Old Soil. A study of the land reform process in Western Wollega, Ethiopia 1975-76*. 90 pp. Uppsala 1977. SEK 45,-. ISBN 91-7106-112-6.
46. Abdel-Rahim, Muddathir, *Changing Patterns of Civilian-Military Relations in the Sudan*. 32 pp. Uppsala 1978. SEK 45,-. ISBN 91-7106-137-1.
47. Jönsson, Lars, *La Révolution Agraire en Algérie. Historique, contenu et problèmes*. 84 pp. Uppsala 1978. SEK 45,-. ISBN 91-7106-145-2.
48. Bhagavan, M.R., *A Critique of "Appropriate" Technology for Underdeveloped Countries*. 56 pp. Uppsala 1979. SEK 45,-. ISBN 91-7106-150-9.
49. Bhagavan, M.R., *Inter-relations Between Technological Choices and Industrial Strategies in Third World Countries*. 79 pp. Uppsala 1979. SEK 45,-. ISBN 91-7106-151-7.
52. Egerö, Bertil, *Colonization and Migration. A summary of border-crossing movements in Tanzania before 1967*. 45 pp. Uppsala 1979. SEK 45,-. ISBN 91-7106-159-2.
56. Melander, Göran, *Refugees in Somalia*. 48 pp. Uppsala 1980. SEK 45,-. ISBN 91-7106-169-X.
58. Green, Reginald H., *From Südwestafrika to Namibia. The political economy of transition*. 45 pp. Uppsala 1981. SEK 45,-. ISBN 91-7106-188-6.
59. Isaksen, Jan, *Macro-Economic Management and Bureaucracy: The Case of Botswana*. 53 pp. Uppsala 1981, SEK 45,-. ISBN 91-7106-192-4.

60. Odén, Bertil, *The Macroeconomic Position of Botswana*. 84 pp. Uppsala 1981. SEK 45,-. ISBN 91-7106-193-2.
61. Westerlund, David, *From Socialism to Islam? Notes on Islam as a Political Factor in Contemporary Africa*. 62 pp. Uppsala 1982. SEK 45,-. ISBN 91-7106-203-3.
64. Nobel, Peter, *Refugee Law in the Sudan. With The Re-fugee Conventions and The Regulation of Asylum Act of 1974*. 56 pp. Uppsala 1982. SEK 45,-. ISBN 91-7106-209-2.
66. Kjærby, Finn, *Problems and Contradictions in the Development of Ox-Cultivation in Tanzania*. 164 pp. Uppsala 1983. SEK 50,-. ISBN 91-7106-211-4.
68. Haarl v, Jens, *Labour Regulation and Black Workers' Struggles in South Africa*. 80 pp. Uppsala 1983. SEK 45,-. ISBN 91-7106-213-0.
69. Matshazi, Meshack Jongilanga & Tillfors, Christina, *A Survey of Workers' Education Activities in Zimbabwe, 1980-1981*. 85 pp. Uppsala 1983. SEK 45,-. ISBN 91-7106-217-3.
70. Hedlund, Hans & Lundahl, Mats, *Migration and Social Change in Rural Zambia*. 107 pp. Uppsala 1983. SEK 50,-. ISBN 91-7106-220-3.
71. Gasarasi, Charles P., *The Tripartite Approach to the Resettlement and Integration of Rural Refugees in Tanzania*. 76 pp. Uppsala 1984. SEK 45,-. ISBN 91-7106-222-X.
72. Kameir, El-Wathig & Kursany, Ibrahim, *Corruption as a "Fifth" Factor of Production in the Sudan*. 33 pp. Uppsala 1985. SEK 45,-. ISBN 91-7106-223-8
73. Davies, Robert, *South African Strategy Towards Mozambique in the Post-Nkomati Period. A Critical Analysis of Effects and Implications*. 71 pp. Uppsala 1985. SEK 45,-. ISBN 91-7106-238-6.
74. Bhagavan, M.R. *The Energy Sector in SADCC Countries. Policies, Priorities and Options in the Context of the African Crisis*. 41 pp. Uppsala 1985. SEK 45,-. ISBN 91-7106-240-8.
75. Bhagavan, M.R. *Angola's Political Economy 1975-1985*. 89 pp. Uppsala 1986. SEK 45,-. ISBN 91-7106-248-3.
77. Fadahunsi, Akin, *The Development Process and Technology. A case for a resources based development strategy in Nigeria*. 41 pp. Uppsala 1986. SEK 45,-. ISBN 91-7106-265-3.
78. Suliman, Hassan Sayed, *The Nationalist Movements in the Maghrib.A comparative approach*. 87 pp. Uppsala 1987. SEK 45,-. ISBN 91-7106-266-1.
79. Saasa, Oliver S., *Zambia's Policies towards Foreign Investment. The Case of the Mining and Non-Mining Sectors*. 65 pp. Uppsala 1987. SEK 45,-. ISBN 91-7106-271-8.
80. Andræ, Gunilla and Beckman, Björn, *Industry Goes Farming. The Nigerian Raw Material Crisis and the Case of Textiles and Cotton*. 68 pp. Uppsala 1987. SEK 50,-. ISBN 91-7106-273-4.
81. Lopes, Carlos and Rudebeck, Lars, *The Socialist Ideal in Africa. A debate*. 27 pp. Uppsala 1988. SEK 45,-. ISBN 91-7106-280-7.
82. Hermele, Kenneth, *Land Struggles and Social Differentiation in Southern Mozambique. A case Study of Chokwe, Limpopo 1950-1987*. 68 pp. Uppsala 1988. SEK 50,- ISBN 91-7106-282-3.
83. Smith, Charles David, *Did Colonialism Capture the Peasantry? A Case Study of the Kagera District, Tanzania*34 pp. Uppsala 1989. SEK 45,-. ISBN 91-7106-289-0.
84. Hedlund, Stefan & Lundahl, Mats, *Ideology as a Determinant of Economic Systems: Nyerere and Ujamaa in Tanzania*. 54 pp. Uppsala 1989. SEK 50,-. ISBN 91-7106-291-2.
85. Lindskog, Per & Lundqvist, Jan, *Why Poor Children Stay Sick. The Human Ecology of Child Health and Welfare in Rural Malawi*. 111 pp. Uppsala 1989. SEK 60,-. ISBN 91-7106-284-X.
86. Holmén, Hans, *State, Cooperatives and Development in Africa*.87 pp. Uppsala 1990. SEK 60,-. ISBN 91-7106-300-5.

87. Zetterqvist, Jenny, *Refugees in Botswana in the Light of International Law*. 83 pp. Uppsala 1990. SEK 60,-. ISBN 91-7106-304-8.
88. Rwelamira, Medard, *Refugees in a Chess Game: Reflections on Botswana, Lesotho and Swaziland Refugee Policies*. 63 pp. Uppsala 1990. SEK 60,-. ISBN 91-7106-306-4.
89. Gefu, Jerome O., *Pastoralist Perspectives in Nigeria. The Fulbe of Udubo Grazing Reserve*. 106 pp. Uppsala 1992. SEK 60,-. ISBN 91-7106-324-2.